SOTHIS PRESS

Sothis Press aims to completely reimagine the public domain book with an emphasis on universal access to captivatingly designed print editions. We believe that technology will enable every single book ever published to be perpetually accessible in beautiful print form. Sothis is beauty in ubiquity.

www.sothis.press
1 Waterhouse Square, London, ECIN 2ST

The Spectroscope and Its Applications

J. Norman Lockyer

SOTHIS PRESS

ISBN 978-1-0365-1076-3

The Spectroscope and Its Applications
is in the public domain.

This edition is published by
Sothis Press © 2023. The company
is registered in England and Wales.
The Company Number is 14900389.

NATURE SERIES.

THE SPECTROSCOPE

AND

ITS APPLICATIONS.

BY

J. NORMAN LOCKYER, F.R.S.

WITH COLOURED PLATE AND ILLUSTRATIONS.

London and New York:

MACMILLAN AND CO.

1873.

PREFACE.

THE following Lectures were delivered at the Society of Arts in the year 1869. They have been carefully revised, and in some parts expanded.

I have to thank Professor Roscoe for permission to use several of the woodcuts which appear in his "Spectrum Analysis," to which admirable work the present small volume will, I trust, be found useful as an introduction.

My acknowledgments are also due to Professor Pedler, both for aid in the experimental illustrations when the Lectures were delivered, and for his assistance while the revision has been passing through the press.

J. N. L.

CONTENTS.

LECTURE I.

LECTURE II.

LECTURE III.

DESCRIPTION OF FRONTISPIECE.

1. A Continuous Spectrum, such as is given by the Light from Solid, Liquid, and Densely Gaseous Bodies when in a State of Incandescence.

2—8. Discontinuous Spectra, such as are given by the Light of Gases or Vapours when incandescent.

The examples chosen are :—

2. Sodium vapour.
3. Magnesium vapour, showing the lines due to the rarer and denser vapours.
4. Chloride of strontium, an example of the spectrum of a compound body, the finer lines in the blue-green being alone due to the metallic strontium.
5. Hydrogen at high pressure.
6. Hydrogen at low pressure.
7. Spectrum of a Nebula (Huggins).
8. Spectrum of the Sun's Chromosphere, tracing the bright lines to the radiation of sodium, hydrogen, &c. (Lockyer).

9, 10. Examples of Absorption :—

9. The Solar Spectrum, tracing the dark lines D, C, F, *b*, &c., to the absorption of sodium vapour, hydrogen, magnesium vapour, &c.
10. The absorption of sodium vapour, showing the coincidence of the bright line with the dark line D in the solar spectrum.

LIST OF ILLUSTRATIONS.

THE SPECTROSCOPE

AND ITS

APPLICATIONS.

LECTURE I.

THE field of research which has been opened up by the spectroscope is one with which we have so recently become familiar, that it may almost be said that twenty years ago a course of lectures on the spectroscope would have been an impossibility. The instrument, as we now know it, was then only in embryo; and even at the present time, although immense strides are every day being made, the science of spectroscopy must still be considered in its infancy. And yet, so far as one can see now—it is always very easy to prophesy after the event—there seems very little reason why lectures on the spectroscope should not have been given two centuries ago; for nearly two centuries have elapsed since the immortal Newton made his classical researches on the action of a prism upon sunlight. You may, perhaps, be inclined to ask, how it could take 200 years for a knowledge of the prism, and of the wonders that can be worked by it, to become part and parcel of our common stock of information? If you ask me to explain this, I tell you candidly that I cannot, but there is this grain of caution connected with it which none of us should forget: we may almost say for certain that Newton

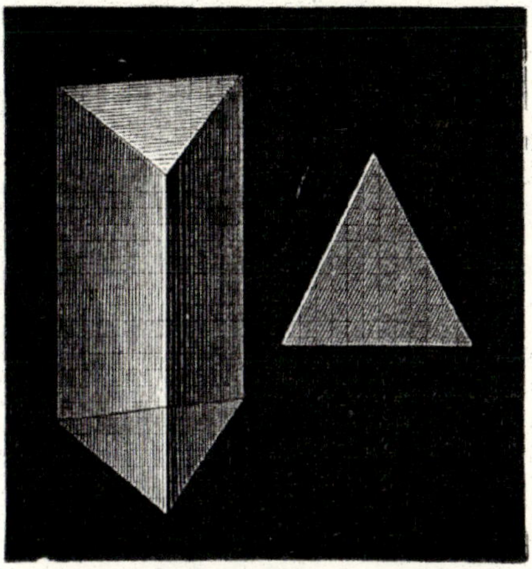

FIG. 1.—Geometrical form of the prism.

and his successors would have brought a great deal more out of the prism than they did, if they had given a little more attention to it, and had tortured it as they did other things. Those who follow us may point to us and say the same; they possibly will say that in the 19th century, men of science, in working and experimenting, saw a great many things, and chronicled them, but did not care to go any further with them. This is very true; and the result is, that work is not done which might be done if we were more receptive and original in our methods of investigation; that is to say, if we trusted Nature more and ourselves less.

I propose that the first part of this lecture should in the main consist of an account of the prism and

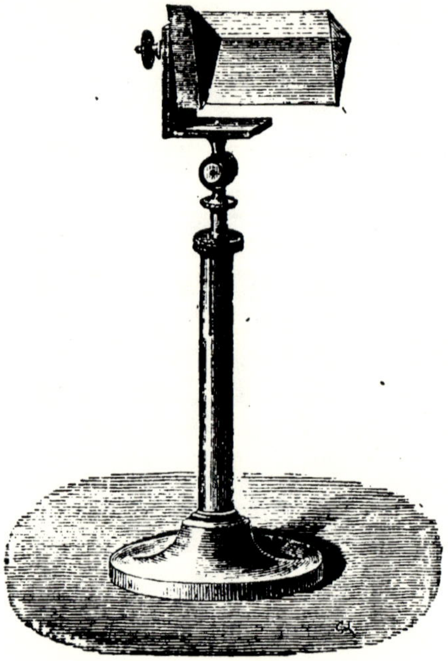

FIG. 2.—Prism mounted on a stand.

the principles of the spectroscope, and then of a description of the various kinds of spectroscopes which are now employed. I hope afterwards to go somewhat in detail into the applications of the spectroscope, not only with regard to terrestrial matters, but also with regard to those problems which we may possibly consider much grander, problems dealing with those celestial bodies which are sufficiently our neighbours to send us light.

Obviously, the first question we have to answer is this, What is a spectroscope? This I answer by saying that a spectroscope is an instrument in which the action of a prism or a combination of prisms is best studied. The next question, then, that arises, is, What is a prism? The accompanying figures (Figs. 1 and 2) will give a good idea of what is meant by a prism, and little time need be spent in description. It is usually a piece of glass—though it need not necessarily be so—bounded by five surfaces, two of which are parallel to each other—though they are not necessarily so—and three of which, bounded by parallel edges, cut each other at different angles; it is in reality shaped like a wedge. The importance of these different angles you will see by and by.

The discoveries of Newton, to which I have already alluded, were connected with prisms, and were based on well-known properties of light.

If a beam of light, as for instance sunlight or an artificial white light, be allowed to enter a dark room from a round hole in a shutter, it will simply travel in a straight line from its source, and to make it deviate from this straight line, one of two things must be done. The beam must either be reflected or refracted.

The reflection of light is of very ordinary occurrence, for when light strikes any polished metallic surface, or in fact a surface of any kind, it is more or less reflected by it. The phenomena of reflection are so well known, the use of the mirror or looking-glass being perhaps one of the most tangible, that no detailed reference need be made to them. The refraction or bending of light takes place when the ray passes obliquely from one medium to another of different density, as from air into water, or from water into air. A simple experiment may be made by passing the beam of light from above into a glass vessel containing water. If the ray strikes the surface perpendicularly, it will be seen that no visible change takes place; the ray simply proceeds directly

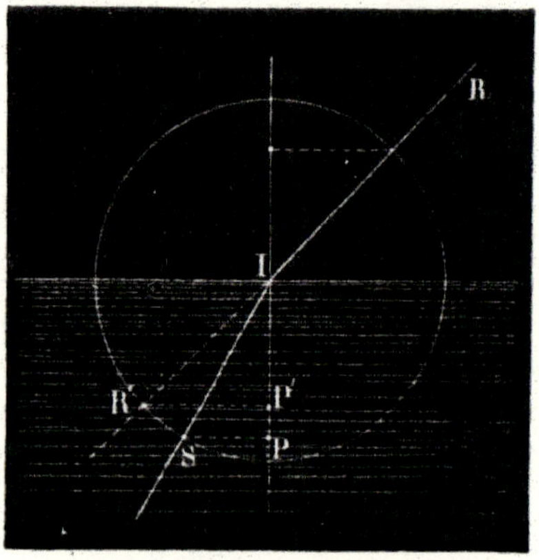

FIG. 3.—Refraction of light.

into the water without altering its direction. If however the beam be allowed to fall on the surface of the water, say at an angle of about 45°, two things may be observed. In the first place a reflection will take place at the surface of the water, that is to say, part of the light will appear thrown back by the surface; and it will be noticed incidentally that the angle at which the reflected ray leaves the water is precisely equal to that at which the incident ray strikes it, thus proving the rule that "the angles of incidence and of reflection are equal;" the second thing to be noticed is, that on entering the water the direction of the beam of light will not be the same as it was in the air. In Fig. 3, the ray R I striking the water at I instead of proceeding to R′ is deflected or refracted to S; that is, the ray will be bent downwards, or, what is the same thing, towards a line I P, perpendicular to the surface, to a definite extent, depending on the angle of the incident ray. The experiment may be varied by allowing the light

FIG. 4.—Explanation of the bent stick.

to fall on the surface at various angles, when it can be shown that the angle formed by the ray, refracted in the water, varies in proportion to the angle of the incident ray, and that the angles formed are bound together by a regular law. Another fact may be observed, that the smaller the angle at which a ray of light strikes the surface of water, or in fact any transparent surface, the greater will be the proportion of light reflected at its surface.

Refraction may be clearly studied by plunging a

stick into a vessel of water: the stick will appear bent at the point where it enters the liquid, as in Fig. 4, thus giving the appearance as if the stick were lifted or bent upwards. Another very instructive experiment is to place a coin at the bottom of a vessel, and then standing so that the coin is just hidden by its edge, to gradually fill the vessel with water, the coin will appear to rise with the bottom of the vessel, and will become visible, as shown in Fig. 5.

The amount of refraction varies with the medium employed, and also with its temperature. The effect of

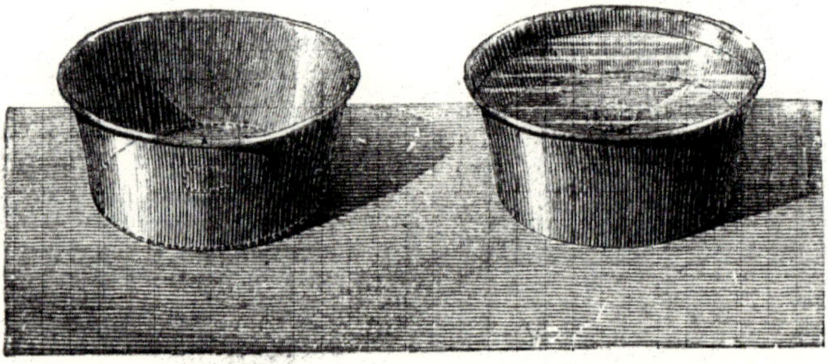

FIG. 5.—Refraction of light. Apparent elevation of the bottoms of vessels.

different media can be clearly seen by passing a ray of light into a vessel, containing a liquid such as bisulphide of carbon, with a layer of water floating on the top. The ray will be seen to be bent on entering the water, and still more bent on passing from the water into the layer of bisulphide of carbon.

We have now to see what takes place when a ray of light enters a piece of glass. We will take first the case of glass with parallel sides. The ray on entering the glass at the upper surface is refracted downwards, as in the case of water, and travels through the glass until it reaches the under surface. Here we have precisely the reverse condition holding,—that is, the ray of light passes from a dense medium to a rarer one; the ray is refracted upwards or away from the perpendicular line, and thus will exactly neutralize the previous refraction, and the beam of light will come out in a direction parallel to its original path, though not quite in the same straight line; as shown in Fig. 6, the ray instead of proceeding in the direction of S′ proceeds in the direction of S.

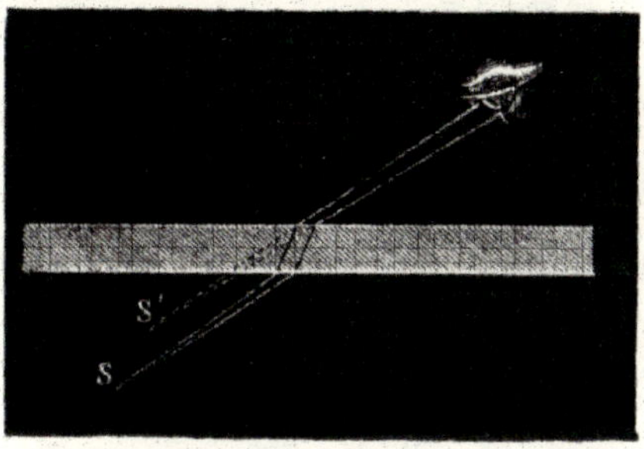

FIG. 6.—Light passing through plate of glass.

If, then, a ray of light passes through a piece of glass, such for instance as a window glass, the surfaces of which are parallel, and inclined to the beam, you see when the beam passes through that the refractive effect is imperceptible. The reason of this is, that when we get the light falling on the glass from the air, then travelling through the glass, and coming into the air again under exactly the same conditions, what is done at the first surface is exactly undone at the second, so that we get pretty much the same effect as at first. But now, if instead of having the glass bounded by parallel surfaces, we use a wedge-shaped piece, or a *prism*, the sides of which are no longer parallel, you will see that there

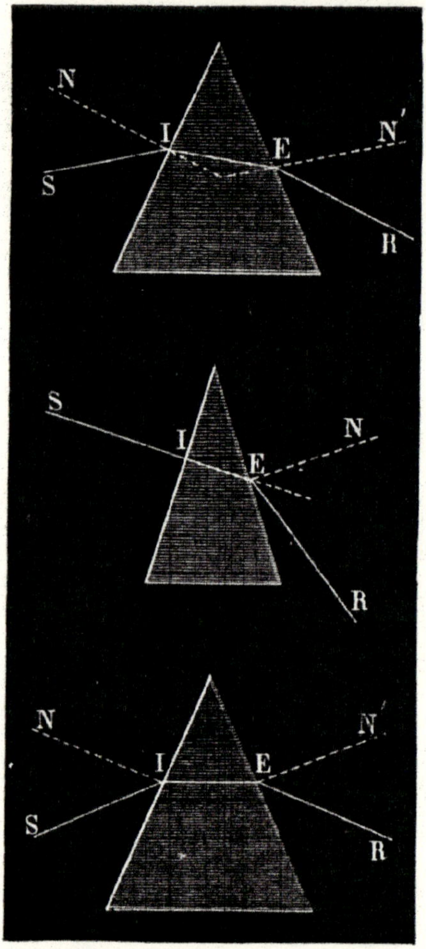

FIG. 7.—Deviation of luminous rays by prisms.

is a distinct alteration in the effect produced; the beam is directed to another portion of the wall altogether. The ray strikes the first side of the prism, and is bent towards the thicker part, or towards a line perpendicular to this surface, and on reaching the second side of the wedge, the ray is

FI . 8.—Images of objects seen through prisms.

again bent in the same direction, towards the base of the prism; for in this case the ray is bent away from the perpendicular to the second surface, and the light emerges from the second surface in a totally new direction. Fig. 7 shows the effect in three cases: the incident ray S I, the path in prism I E, and the refracted ray E R; N I and E N′ being the lines

Fig. 9.—Decomposition of light by the prism. Unequal refrangibility of the colours of the spectrum.

perpendicular to the surfaces. An experiment may easily be tried, which will confirm this. Let a triangular piece of glass be held, with a refracting edge at top, between the eye and a lighted candle, as shown in Fig. 8 ; it will be found that the candle cannot be seen ; but if the prism be gradually raised, the image of the candle will appear, the amount the prism will have to be raised depending on its angle. Now, we have here obtained a deviation or refraction of light, that is to say, it has been bent out of its course ; for we have to look upwards to see the candle. Another effect has also been produced : the light which was white on entering the prism, is now made up of several colours, which are separated more or less from each other ; the candle as seen in the last experiment is not white, but is fringed round with colours. If we again take our beam of light in the dark room as in Fig. 9, and allow it to strike on one of our prisms, so placed that its edges are horizontal, and also that the beam enters it obliquely by one of its surfaces, and then receive the image on a screen, we see a band of colours which reminds us strongly of the rainbow: the lowest colour, if the base of the prism be upwards, will be red, next above orange, passing by imperceptible gradations to yellow, and afterwards green, which then passes through the shades of greenish blue till it becomes a pure blue, then indigo, and finally ends with a violet colour. The transition from one colour to another is not abrupt, but is made in an imperceptible manner, so that it can scarcely be said, for instance, where the yellow ends or the green begins. The cause of this band of colours, or *spectrum* as it is called, was first discovered by Sir Isaac Newton, who tortured this spectrum in several ways. He took one of the colours thus produced, say red, as is shown in the figure, and made it pass through a second prism, receiving the image on a second screen; the image is found to be rather longer, but the colour remains unaltered. This experiment proves that this colour of the spectrum is simple, and the same has been found of all the others. As Newton in his experiment operated with sunlight, the band of colours was in this case called the *solar spectrum.* The rainbow itself is also in reality nothing more nor less than a solar spectrum, which is caused by refraction in the rain drops.

If, instead of getting one beam of white light, we take two of differently coloured lights, red and blue, and pass these two beams of different colour through the same prism, you will see that the action of the prism on these two differently coloured beams will be unequal; in other words, you will get the red beam deflected to a certain distance from a straight line, and the blue deflected to a certain other distance. You see by this experiment that there is a distinct difference in the amount of refrangibility—that the red light is not diverted so far out of its original direction by the prism as the blue. And this leads us to Newton's first proposition, which is this:—"*Lights which differ in colour differ in refrangibility.*" I think that requires no explanation. You will be able to translate it for yourselves thus. Lights which differ in colour are differently acted upon by a prism, which, as you have seen, gives us a considerable result of the action of refraction.

We now approach Newton's great discovery, which is this:—"*The light of the sun consists of rays differently refrangible;*" that is to say, if instead of using two coloured beams as in the last experiment we take a beam of sunlight, and make it pass through a prism, the action of the prism is at once to turn that beam into a beautifully coloured band, which will remind you of a rainbow. It was this which Newton did in a dark room, which led him to

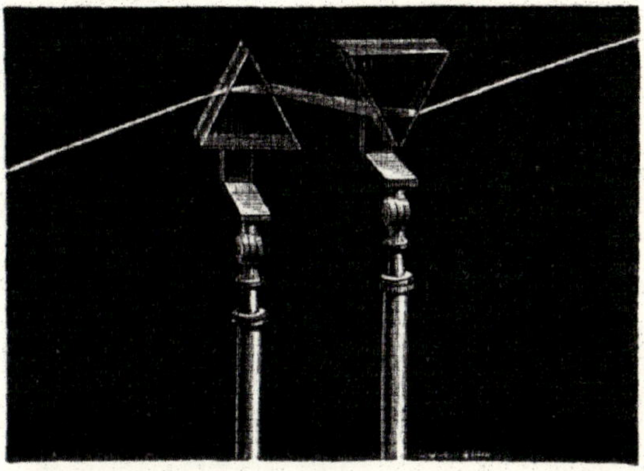

FIG. 10.—Recomposition of white light by means of a second prism.

his important discovery. White light is compounded, of light of different degrees of refrangibility. But how is it possible to show the truth of Newton's assertion that white light is compounded of these different colours? We can do so by simply placing in the path of the coloured beam which you see passing through the room, another prism placed in a contrary direction, as shown in Fig. 10; you see in a moment that we get back white light; for the second prism exactly neutralizes the effect caused by the first, and the ray proceeds as if nothing had happened.

Possibly you may ask, is it true that white light is built up of all colours? That question can be answered to a certain extent by an experiment of a different order. If a disc, divided into sections and coloured with the principal colours of the spectrum,

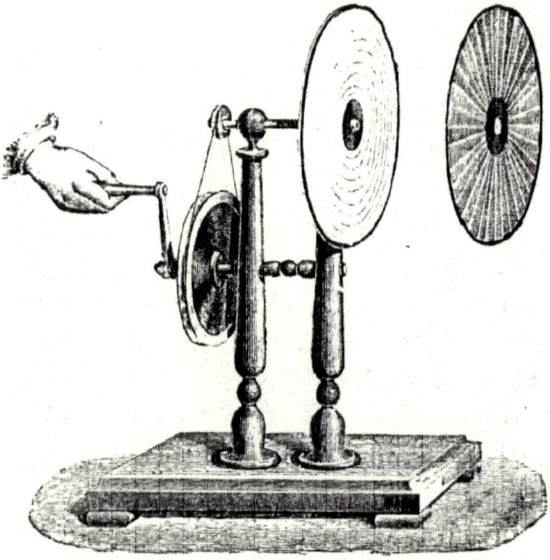

FIG. 11.—Recomposition of white light by means of a rapidly revolving disc, coloured in sectors.

as shown in Fig. 11, be taken, and if it be true that the idea of white light is simply an idea built up by the eye, because we have all these multitudes of light waves perpetually pouring into it with a velocity that is very much greater than anything which can be translated into words, surely we should get something like this effect also if we were able, by rapidly rotating this screen, to obtain a more or less perfect substitute for white light. The coloured disc being made to rotate rapidly, you see we obtain something like an approximation to white light, though the white colour does not come out so clearly as it might do. Now I am very anxious that you should see that this is really an effect due to the flowing in of light from different parts of that wheel into the eye, and so forming this compound impression, which is conveyed to the brain; and so if instead of illuminating the disc continuously by the electric lamp, or by sunlight, it is illuminated intermittently, by an electric spark, you would see that, although the disc is rotating rapidly all the time, each separate colour is now discernible, and the disc appears to stand still. The reason of this difference is, that in one case the rotation of the wheel builds up a compound image in the eye, and in the other case it cannot do so, because the flash of the light is much more rapid and instantaneous than the rotation of the wheel.

There is one more experiment which can be easily made, to show that all the beautiful colour which we get in nature is really reflected after all, and that if our sunlight, instead of being polychromatic—that is to say, compounded of all these beautiful colours—were monochromatic, or of one colour only, the whole expanse of creation would put on a very different appearance from what it does. If, instead of illuminating a diagram, the letters of which are of different bright colours, by the white light of the electric lamp, we illuminate it by a light that only contains one colour—by the yellow light of sodium, for instance—and then look at the diagram, you will see that some of the letters upon it are almost invisible, whilst others are very clear, the yellow light only allowing a difference to be seen of more or less depth of shade—there being no difference in colour. But when we allow the polychromatic light from the lamp, or as we get it from the sun, to shine upon the diagram, you at once see that all these letters are of different colours, and burst out, as it were, into beauty. This experiment feebly indicates the advantage we possess in living in a universe lit by white or polychromatic light, instead of light which is merely blue, or yellow, or any other single colour.

Hitherto we have spoken only of refraction. I now introduce the word *dispersion,* which represents simply a measure of different refractions, or the difference between the bending of the red and the violet rays of light. In an ordinary spectrum, the difference between the red and the violet is the difference of the refraction of those two colours by the prism, and the angle which the red, or yellow, or other colour, forms with the original path of the compound-beam is called the *angle of deviation.*

There is one other consideration which we owe to Newton. In his very first experiments, that great philosopher discovered that the quality of the spectrum depended very much on the following consideration:—If I wish to get the best possible effect out of a prism, and the purest possible spectrum, I have so to arrange it that the particular ray which I wish to observe, whether the yellow, the blue, the green, or any other, leaves that prism at exactly the same angle as the incident compound ray falls on it. This angle is termed the *angle of minimum deviation.*

The two things, therefore, of greatest importance in this subject which we owe to Newton, are, first, the explanation of the dispersive power of the prism; and, next, the pointing out the extreme importance of arranging the prism, so that if we want to observe any particular part of the spectrum, the rays constituting that part of the spectrum should leave the prism at the same angle as the white light falls on it.

It is very curious, however, that Newton, although he made many experiments on prisms, really omitted one of the most important points, which you will see carefullv arranged for in every one of the spectroscopes used at the present day. And here again we get an idea of the enormous patience which is necessary in these matters, for we had to wait a century and a quarter before the next essential point was hit upon in the construction of a spectroscope. Newton made a round hole in a shutter for his experiments, but we now know he ought not to have done that; he ought to have made a slit. But this did not come out until 1802, when Dr. Wollaston, by merely using a slit instead of a round hole, made a tremendous step in advance. You will see the importance of this in a moment. If we take a cylindrical beam of sunlight and put a prism in the path of the beam, we observe that the spectrum is not a pure one; but if we change the round hole for a slit, we obtain a spectrum of the greatest purity; the red, blue, green, and violet, instead of overlapping and destroying the beauty of the spectrum, show distinctly as simple colours, each one speaking for itself on the screen. By using this narrow slit instead of the round hole which Newton made in the shutter, we got the first idea of the tremendous importance of spectrum analysis; for no sooner had Dr. Wollaston examined the sunlight with the new arrangement, as Newton had done a century and a quarter before with the old one, than he found out that it was not at all as Newton had represented it. Newton told us in fact that the sunlight was continuous, that is to say, that the spectrum was one in which there was no break in the light which flowed out to every part of the spectrum, from the extreme red to the violet. When Dr. Wollaston tried the slit, he found, however, that the spectrum, instead of being that rainbow band of light which you have seen, was really broken by a succession of fine—beautifully fine—black lines.

These lines were observed by Dr. Wollaston, but it was not until 1814 that we find them mapped out with the greatest care, to the number of 576, by a German optician, named Fraunhofer; hence they are termed "Fraunhofer lines," the principal ones being lettered A, B, C, &c.

If we say, then, that spectroscopic inquiry dawned with Newton, certainly the sun began to rise with Fraunhofer; for he, no longer content with getting a sunbeam through this slit, and finding out and measuring with most admirable accuracy these 576 lines in that band of colour, turned his telescope to the moon and the planets, and the different stars; and he discovered that, in the case of the stars,

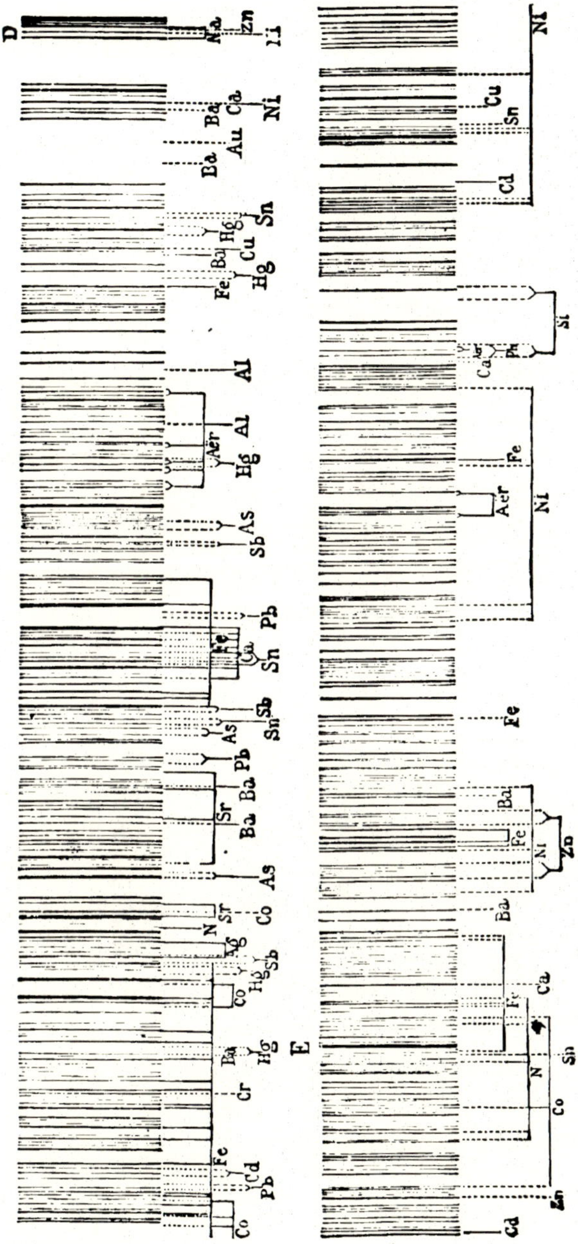

FIG. 12.—Left-hand diagram, Solar Spectrum near D; right-hand diagram, Solar Spectrum near E.

the positions of the lines varied considerably from those they occupied in the spectrum of the sun; and this is one of the most important discoveries which has been made during the present century in these matters. Indeed, it is the foundation of very much of the later more detailed work.

The solar spectrum, then, as we have said, far from being continuous, is crossed by an almost innumerable number of dark lines, some being fine and others thicker and blacker. Fig. 12 shows a small portion of the spectrum in the yellow and green. Other observers, such as Kirchhoff, Thalén, and Ångström, have worked at these dark lines, and have drawn most beautiful and elaborate maps, showing at least 2,000 lines of various thicknesses.

We have now to pass on from 1812 to the year 1830, when Mr. Simms, an optician of world-wide reputation, made another very important improvement in the spectroscope. Instead of merely using a prism and observing the slit with the naked eye, he placed a lens in front of the prism, so arranged that the slit was in the focus of the lens. The light which is allowed to pass through the slit is thus turned into a cylindrical beam and thus travels through the prism; then, instead of having merely the eye to observe the spectrum, there is another lens which grasps the circular beam and compels it to throw an image of the slit, which may be magnified at pleasure for each ray. The great importance of this is at once obvious, if you think for one moment of the figure showing the lines in the solar spectrum. We now know, and it is not too early to place this before you, that these black lines indicate regions in the spectrum where there is no light. If the light is perfectly continuous, so that every ray of light is enabled to register itself at the end of the telescope, by painting an image of the slit, you will get a continuous spectrum; but supposing, for instance, that the whole of the yellow light were absent, it is clear that the spectroscope, if it does its duty well, will give you blackness where the yellow light is absent. We do not find that the whole of any particular colour is absent, but here and there, scattered over all the colours, there are places where the rays of light do not come to tell their story. This is the explanation of the Fraunhofer lines in the solar spectrum. In the light which we get from the sun, certain of the rays which we may suppose ought to come to us, do not come, and we get no news from them. We do get news of some of the other rays, which show us the various shades of blue, of green, and so on; but here and there a ray, which possibly might have come if it were not better employed, does not come, and therefore the image of the slit cannot be painted. I am glad to say that we know a little more about these lines than we did some years ago. You may imagine the enormous mystery—the wonderful reverence almost—with which this question of the Fraunhofer lines was approached, until they were thoroughly understood; and recollect that we owe the discovery of them—by which we are enabled now to determine the pressures acting in the atmospheres of the most distant stars—simply to the fact that Dr. Wollaston, instead of drilling a round hole, used a slit, and to the other additional fact, that Mr. Simms, instead of using that slit with a mere prism, used a lens and made the beam parallel, and then allowed that parallel beam, after it had passed through the prism, to pass into another telescope, and form an image of the slit for each ray. You see how closely connected are the grandest discoveries with the skill and suggestiveness of those who supply different instruments for our use.

Now I must ask you to come back again to the prism. I have already told you that dispersion is the measure of the difference of the refrangibilities. If we take a prism which appears like an ordinary one, but really is composed of several layers of different kinds of glass, and pass an ordinary beam of light through it, it will be differently acted upon by the various layers, and we shall get a difference in the spectra. We have here in fact three distinct spectra, showing that there is something in the different layers of which this prism is composed which turns the light out of its path, and which disperses it more in some cases than it does in others. The cause of this is the density of the glass composing each layer: some kinds of glass are nearly twice as heavy as others, and fortunately we are not limited to glass, for if we were we should not be able to go so far in these inquiries as we do. The prism I am using consists of three separate pieces of glass of different density, and it may be seen that the three spectra obtained are differently refracted. It is a very natural conclusion that the heavier and denser glass should have a stronger action on the light than the lighter glass has. So that, in these inquiries, if we want to get great dispersion, not only must we use heavy glass, but we leave glass behind altogether, as amongst the liquids we find some which give even a greater dispersion than the densest glass. If a beam is passed through a hollow prism of glass filled with bisulphide of carbon, the spectrum obtained is much longer than that produced by the densest flint glass we can get. But there is another consideration to be borne in mind. The dispersive power and refractive power not only depend upon the density of the glass, but on the refracting angle of the prism. If a beam of light is sent through two prisms of unequal angles the effect is unequal. Thus, if we take one prism with an angle of 20°, and another with an angle of 60°, the larger angle gives us a greater deviation and dispersion; therefore, we not only have density to help us, but we have also the angle of the prism.

And now let us go on to a third important point in the matter. We are not limited to one prism, if we wish to get a great amount of dispersion; if you will think the matter over, you will see that there is no good reason why we should not employ two, and then you will find that the dispersion will be considerable. So you see, first, we have a single prism of a dense substance; by increasing the angle we get increased dispersion, and then we get it still further increased by adding another prism, and so we might go on, adding prism after prism, until we get to any number of prisms arranged in the best possible manner for the light to be successively dispersed by each of them. First of all, you have the dispersive power of glass, then you have the angle of the prism, and then you have a number of prisms, all of them capable of being so arranged that we can make them all useful in these inquiries, until at last we get a dispersion of such an enormous amount, that the spectrum of the sun, as mapped by Kirchhoff and Bunsen, is some yards in length, although it is nothing but a succession of images of one of the finest slits which our best opticians are able to make.

You see, therefore, that our spectroscope depends first of all on Newton's work with the prism in 1675, and on the fact which Newton found out incidentally, that it is important that the prism should be used at the angle of minimum deviation. We then get the slit added by Wollaston in 1812; then the collimating lens added by Simms, in 1830. In this way we have arrived at the spectroscope improved and modified as an instrument, until at last we get spectroscopes so arranged that the glass is of the finest possible material, the angle the largest possible, the glass the densest possible, and the number of prisms as great as possible.

There are some other considerations connected with the manufacture of spectroscopes which it is hardly necessary I should bring before you, as they are rather more in the nature of detail than of general principles; but I must point out that where liquids are employed, it is absolutely essential that the temperature should be as equable as we can get it. A current of warm air in a room is quite sufficient to render any spectrum obtained by these liquid prisms perfectly useless; hence, although their great dispersive power is of great value in some cases, where we want dispersion more than anything else, still, as a rule, we are limited for nearly all our researches to these dense glass prisms of great angle, to which I have already alluded. But

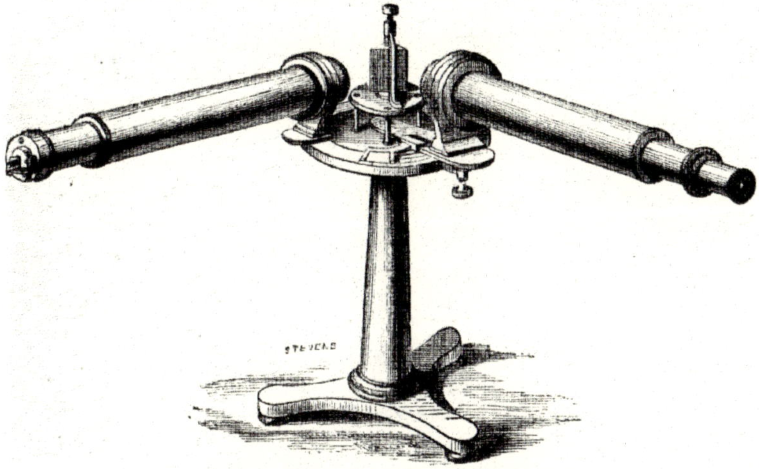

FIG. 13.—Chemical or students' spectroscope

there is another consideration of great importance which comes in here. If the angle of a prism be large, a ray of light travelling from one prism to another enters the second at an extremely small angle, under which circumstances a large amount of light is reflected, but still it is not better to use a greater number of prisms of a smaller angle than a smaller number of a larger one. Again, in spectroscopes of many prisms it is essential that there should be some arrangement by which each part of the spectrum should be observed with each prism at the angle of minimum deviation for that ray. This may be done in many ways, and the beam may be made to pass back again through the prisms, thus doubling the amount of dispersion. On these points I shall have more to say presently. Another important consideration, besides the purity of the

FIG. 14.– Spectroscope with two prisms.

material, is the perfect figure of the slit. You might imagine that the slit of a spectroscope was perfectly easy to make; but, judging by the results of the manufacture, it is extremely difficult, for a perfect slit is still very rare, the best being made by Steinheil of Munich. Mr. Browning has suggested making the slit of a compound of gold, which will not rust, or be acted upon much by temperature, and which also will take a good figure without any very great difficulty.

So far, I have spoken of spectroscopes as spectro-

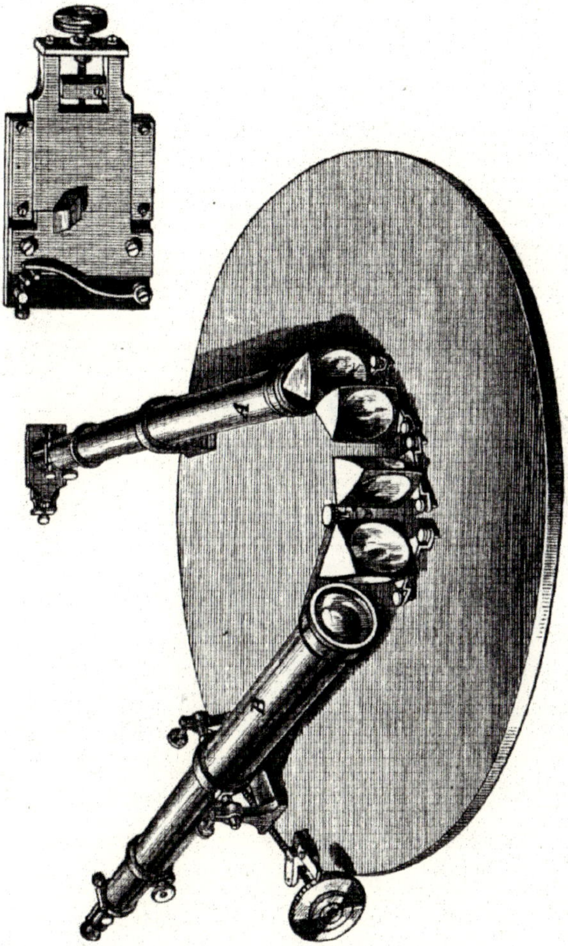

FIG. 15.—Steinheil's form of four-prism spectroscope: arrangement of slit shown separately.

scopes—as one of the instruments the improvement of which should be cared for by every student in science. Their applications will come after. As may be imagined, spectroscopes are now constructed with one, two, three, four, or more prisms, the number depending on the purpose for which they are to

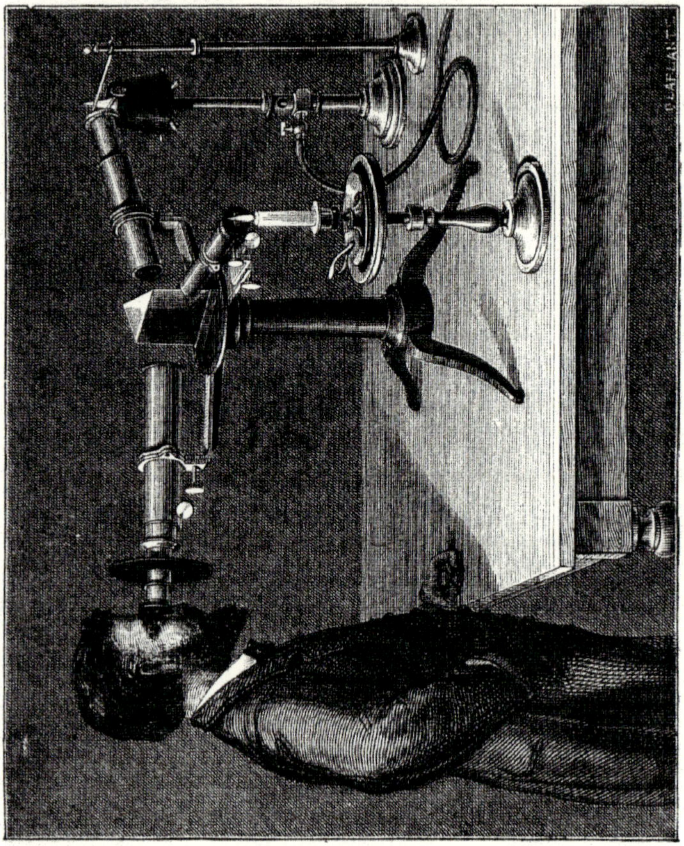

FIG. 16.—Spectroscope with reflected scale.

be employed. The instrument shown in Fig. 13 may be called a chemical spectroscope, for an instrument of this kind is now almost as important and essential in a chemical laboratory as a balance. Spectroscopes are also constructed with two prisms, as shown in Fig. 14; these are used in cases when rather more dispersion is desired than can be obtained with the one-prism instrument. When, however, any accurate and elaborate work has to be done, such as in carrying out original investigations, more prisms have to be employed. The engraving given in Fig. 15 is of an instrument which historically is extremely interesting, as being the one with which Kirchhoff made his most elaborate and accurate maps of the solar spectrum; it is furnished with a battery of four large prisms, which give a great deviation and dispersion. There is no reason why spectroscopes of many more prisms should not be employed, except that they require to be worked only with strong lights, as light is here so much dispersed or spread out that a feeble spectrum would be almost lost.

As the principle of construction is almost the same in all kinds of spectroscopes, we had better commence by a description of the simplest form, namely that with one prism, as shown in Fig. 16. It will be seen to consist of a circular table, supported by a pillar and three legs, carrying three lateral tubes; the right-hand tube is called the collimator, and holds at its outer extremity the fine slit, the width of which can be regulated to a nicety by a micrometer screw; the other end of the collimator is furnished with a lens, which serves to collect the rays of light coming from the slit, and to render them parallel before falling on the prism in the centre of the table. The prism is so placed and fixed by a clamp that the light entering the slit from the source of light, shown in the figure as a gas lamp, strikes it and leaves it at what is called the *angle of minimum deviation*, a term which has already been explained; after passing through the prism, in which the light undergoes both deviation and dispersion, the spectrum

FIG. 17.—Huggins's star spectroscope.

is observed by the telescope on the left, which is simply a small astronomical telescope of low magnifying power. There are two methods of measuring spectra. The telescope may be attached to a movable arm, which can be directed to any part of the spectrum that may be required; and the outer edge of the circle along which the telescope moves, may be graduated with an accurate scale of degrees, which can be divided with more or less minuteness, according to the precision in the exact position of the dark lines, &c. required in various spectra. In this method the line to be measured is brought into the centre of the field of view of the observing telescope, and the position of the telescope read off. Of course if the line measured is situated in the red end of the spectrum, the telescope will be in a different position from that it would occupy if the line were in the blue end. The second method of measurement may be gathered from Fig. 16. It consists of a short tube carrying at its outer extremity a small photographic scale, which is illuminated by a candle flame; the light passing from the photographic scale is rendered parallel and thrown on the surface of the prism by means of a lens in the tube, carrying the scale, and is reflected by the last surface of the prism up the observing telescope, so that it is seen as a bright scale on the background formed by the spectrum under observation.

The spectroscope has also been adapted to the telescope with very great success; for it is essential not only to determine the spectra of the light emitted by various substances in our laboratories on this earth, but also the different spectra and positions of the dark lines or bright ones, obtained from the various orders of celestial objects, such as the sun and stars, comets, nebulæ, planets, and so on; we must for this purpose have something attached to the telescope. Fig. 17 shows a star spectroscope, which differs in arrangement only and not in principle from other spectroscopes, except in one point to which I have to draw attention with regard to this spectroscope. I have insisted on the importance of the slit; but you will see in a moment that the image of a star, if it is a good image, will be a mere point in the telescope, and therefore, while a slit is not absolutely necessary, it is essential to have some arrangement by which that point of light, the spectrum of which would be merely a line, and therefore not broad enough to enable us to see what the lines are which we may expect in the spectra of stars (if they be anything like the spectrum of the sun), shall be turned into a band. That has been accomplished by means of a cylindrical lens, its function being to leave the light alone in one direction, but to turn it into a band in another direction, so that when the light of the star gets through such a lens, it is no longer a point but a line, and this is then grasped by the collimating lens, sent through the prisms, and received by the observing telescope, so that when you get the image of it in the observing telescope, instead of having a line of light so fine that the lines in it cannot be distinguished, it is a distinctly broad band in which the lines can be observed. As this lens is simply a contrivance for enabling the eye to observe the lines, I submit now, as I submitted some years ago, that a good place for it is close to the eye, between the eye and the image. I have been gratified to find that, in many of the spectroscopes used on the Continent, this arrangement is adopted.

We have now an idea of the action of the simple prism. I will next bring to your notice another kind of prism, which differs from the simple one very much as the achromatic telescope differs from the non-achromatic one, which was the first attempt made at an instrument for astronomical observations. Many of you know that the object-glass of a telescope, as now constructed, consists of two lenses made of different kinds of glass. Of course, we have dispersion and deviation at work in both these kinds of glass, but the lenses are so arranged, and their curves are so

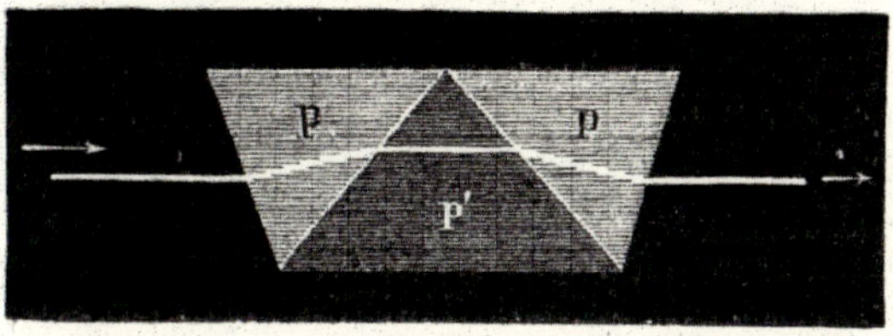

FIG. 18.—Direct-vision prism with three prisms, shewing path of ray.

chosen, that, as a total result, the deviation is kept while the dispersion is eliminated, so that, in the telescope, we have a nearly white image of anything which gives us ordinary light, although, as you know, it is by the deviation alone that we are enabled to get the magnified image of that object. So also in the spectroscope we have an opportunity of varying the deviation and the dispersion. By a converse arrangement we can keep the dispersion while we lose the deviation; in other words, we have what is called a direct-vision spectroscope. If we take one composed of two prisms of one kind of glass which possesses a considerable refractive power, and three prisms of another kind which does not refract so strongly, arranged with their bases the opposite way, the deviation caused by the two prisms in the one direction will be neutralized by the deviation of the three prisms in the opposite direction; whilst the dispersion by the three prisms, exceeds that which is caused by the two prisms in the opposite direction, the latter dispersion, therefore, will neutralise a portion only of the dispersion due to the three prisms. The final result is that there is an outstanding dispersion after the deviation has been neutralized, so that when we want to examine the spectrum of an object we

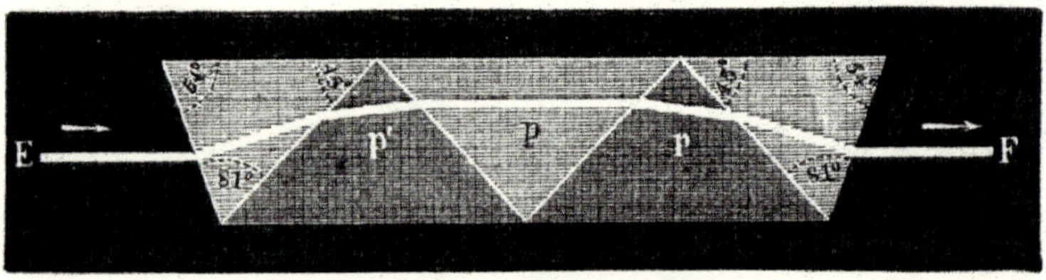

FIG. 19.—Direct-vision prism with five prisms.

no longer have to look at it at an angle. No doubt you recollect the angle that was made by the light the moment it left the prism, but we have an opportunity, by this arrangement, of seeing the spectrum of an object by looking straight at the source of light. In the application of spectrum analysis, especially to the microscope and telescope, this modification—due to M. Janssen, the well-known astronomer, who was the first to bring it into general notice—is one of great practical importance, so that in any research which does not require excessive dispersion, this direct-vision arrangement is getting into common use. I have here another direct-vision arrangement which is well worthy of being brought to your notice. It does not depend at all upon the principles I have just been trying to explain to you. It is called the Herschel-Browning direct-vision spectroscope, in which the ray is refracted and reflected internally, in the prisms themselves.

We have therefore, in addition to the simple prism which I formerly brought to your notice, two other aids to research of extreme value in certain classes of observations. The direct-vision spectroscopes which are now sold are made on one of the two principles just described; some of them are made so small that they can be easily carried in the waist-coat-pocket, and still are so powerful that all the principal, and many of the less prominent, lines in the solar spectrum may be seen with them.

Of the special application of the spectroscope to the microscope I need say but little now. The spectroscope thus used is a direct-vision one, this form being far more convenient for attaching to the microscope. The light illuminating the object in the microscope was first of all passed through a prism; but in later arrangements it passes through the prism in its passage from the object. This is obviously a much better plan, because, in the first instance, you could only deal with transparent objects; but here, as you deal in any case with the light that comes from the object itself, it is quite immaterial whether the object be opaque or transparent.

LECTURE II.

IN what has now been stated we first saw Newton founding spectral analysis, by using a hole in a shutter and a prism; then we discussed Wollaston's substitution of the slit: after that Mr. Simms' introduction of the collimating lens was referred to; and then the growth of the modern spectroscope.

It is time, now, that we came to the applications of the instrument. And in dealing with these applications I shall divide my subject into two perfectly distinct portions. I shall first deal with those which depend upon the different modes in which light is given out or radiated by various bodies under different physical conditions—with, in fact, the *radiation* of light. And, in the second place, I shall deal with the spectroscope's story of the way in which white light giving a continuous spectrum is stopped or absorbed by different transparent bodies—with, in fact, the *absorption* of light.

The first application of this question of radiation is one of the most general importance. It enables us to differentiate between solid, liquid, and gaseous substances, and between gaseous or vaporous substances in different stages of pressure. If, for instance,

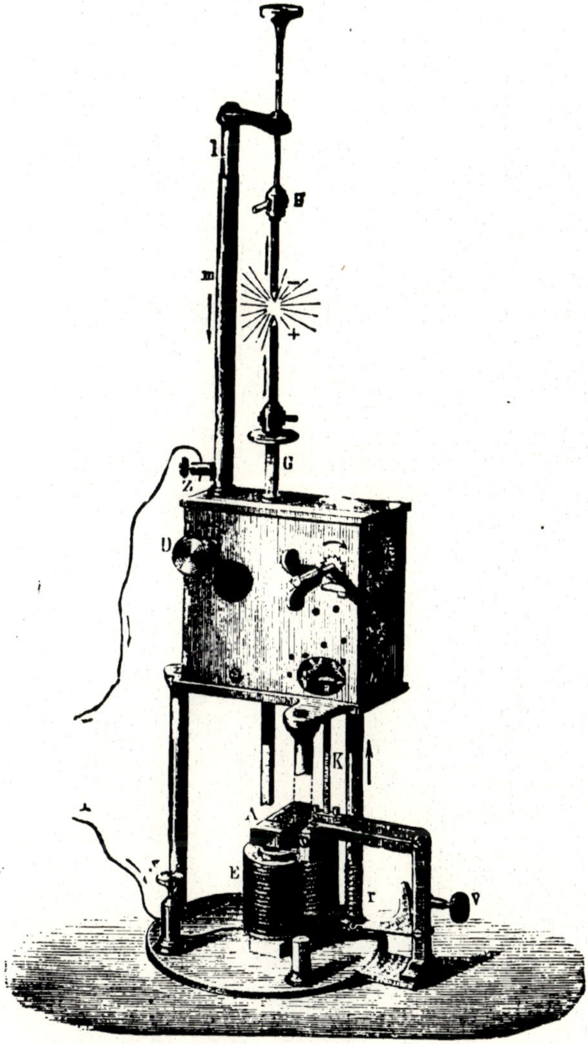

FIG. 20.—Electric lamp.

we take a platinum wire and heat it to redness, and examine by means of the spectroscope the light emitted, we shall find that only red rays are visible; then if the wire be gradually heated more strongly, the yellow, green, and blue rays will become visible,

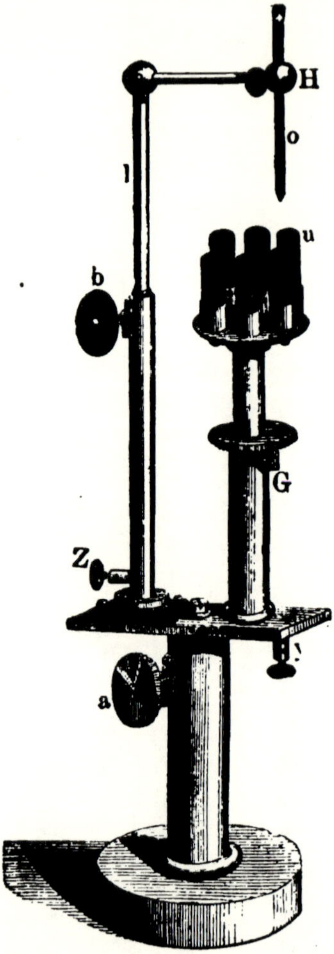

FIG. 21.—Arrangement of the electric lamp used for rapid comparisons.

until finally, when the wire has attained a brilliant white heat, the whole of the colours of the spectrum will be present. If I were to burn a piece of paper, or a match, or ordinary coal-gas, you all know we should get a white light, but you may possibly not all know that if we raise any solid or liquid to a state of incandescence or glowing heat we should get exactly that same sort of light, which will always give us a continuous spectrum. Before a large audience the best method of showing this fact is to use an apparatus called the electric lamp (Fig. 20), and to pass the current of electricity through two carbon points, which are intensely heated by their resistance to the passage of the current. The spectrum obtained from these points, by means of the dispersion of two bisulphide of carbon prisms, is quite continuous from end to end. Now carbon is a solid, and therefore if we take carbon as an example of a solid or liquid substance in a state of vivid incandescence, and we obtain from these carbon points a continuous spectrum, you must accept that as an indication of the truth of what I say, for I have not time to experiment on every solid and every liquid substance. The spectrum is received on the screen, and you see it is continuous, that is to say, there are no breaks, such as those we saw in the figure representing a portion of the solar spectrum on page 20, where the black lines represent the breaks in the solar spectrum which are called the Fraunhofer lines.

Let us then consider this fact established, namely, that solid or liquid bodies, when heated to a vivid incandescence, give a continuous spectrum without bright lines. Under these circumstances the light to the eye, without the spectroscope, will be white, like that of a white-hot poker; if the degree of incandescence is not so high, the light may only be red, like that of a red-hot poker. But so far as the spectrum goes—and it will expand towards the violet, as the incandescence increases, as before stated—it will be continuous.

Now, suppose, instead of giving you the spectrum of these solid white-light-giving carbon points or that from an ordinary gas flame, I show you the spectrum of a light source which is coloured. If, for instance, we burn some coloured fire, such as the red fire of our pyrotechnic displays. You must not consider that this is sensational, for Sir John Herschel, very many years ago, was on the eve of discovering the great point of spectrum analysis which I have to bring before you, by merely examining these coloured fires. If we examined such a light by means of the spectroscope, you might expect that we should obtain the red end of the continuous spectrum; that on burning green fire we should see the green portion of the spectrum and so on. But this is not so; we find that the background of the spectrum is dark or nearly so, and that we have certain localizations of light, or bright lines, in different parts of the spectrum. Now, the differences in colour are accompanied by differences in the spectra. We have something very different from the continuous spectrum we had before, and this is, in fact, one of the first practical outcomes of spectrum analysis. It enables you in a moment to determine the difference between a solid or liquid body, which gives you a continuous spectrum, and a vapour or gas, which gives you a spectrum containing bright lines. The reason that different vapours and gases are of different colours is now clear; if we examine the light by means of a spectroscope, we find that the light rays which they emit are located in different parts of the spectrum.

In these instances, then, the spectra consist of lines which are located in different parts of the spectrum. Let us burn some sodium in air, and then

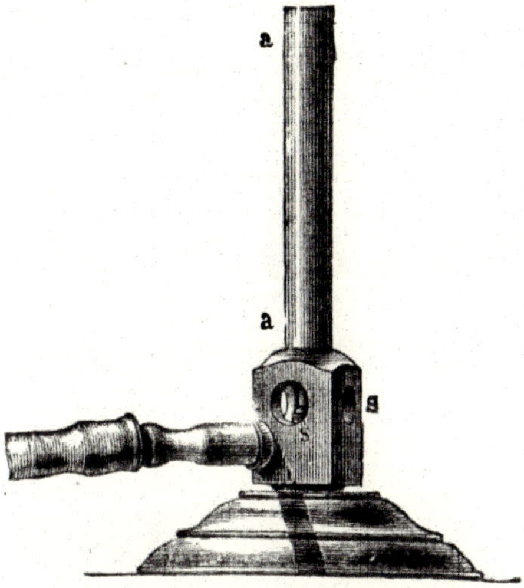

FIG. 22.—Bunsen's burner for flame spectra.

examine the spectrum of its vapour, or better still, let us place some sodium, or a salt of this metal, such as table salt, in a gas flame which is consuming a mixture of air and gas, in a burner known under the name of a Bunsen's burner (Fig. 22), the bluish flame of which is due to the complete combustion due to the greater supply of air from the holes at the bottom. The flame immediately becomes of an intense yellow colour due to the vapour of sodium. In this we have further evidence of the connection between the colour of the light which we get from a vapour and the spectrum of that vapour. It is usual to place the salt to be examined in a platinum spoon, and insert it in the flame; but the utmost constancy is insured by adopting an arrangement of Mitscherlich's, shown in the accom-

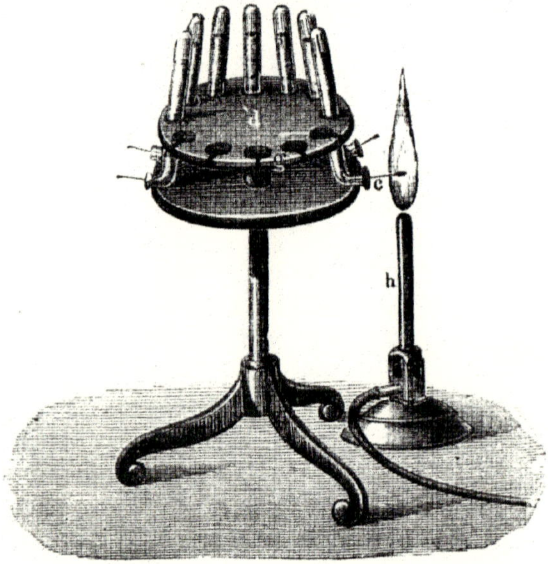

FIG. 23.—Mitscherlich's arrangement for flame spectra.

panying drawing (Fig. 23), in which a platinum wick is kept continually moistened by a solution of the salt, generally the chloride, the spectrum of which is required to be examined. You will imagine, *à priori*, from what I have already said, that as in the case of sodium vapour, the colour of the light is orange, the line of the vapour will appear in the yellow or orange part of the spectrum, and you will not be mistaken. For you will see on examining this flame with a spectroscope, that we obtain a spectrum consisting of a brilliant yellow line upon an almost black background; if, however, the flame is observed by means of a very narrow slit, this line will appear double, that is, it really

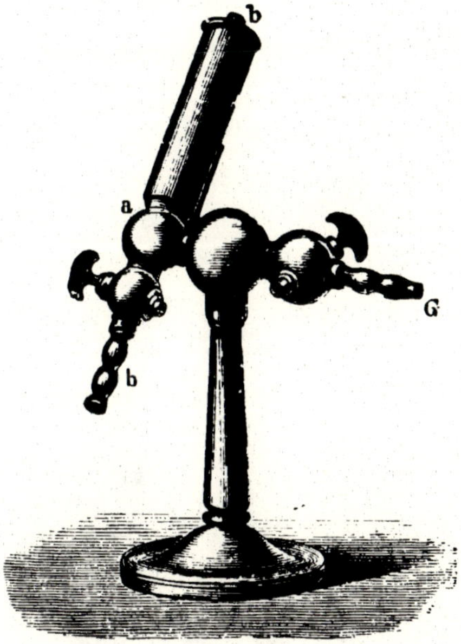

FIG. 24.—Herapath's Blow-pipe.

consists of two extremely fine lines which are very close to each other, and if the slit be wide the images overlap one another.

If we then pass on to another substance, and take some lithium instead of sodium, we obtain a brilliant carmine tinted flame, which on examination by the spectroscope is found to give a spectrum consisting of one splendid red, and a fainter orange line. Potassium gives a violet-coloured flame, and yields in the spectroscope a red line and a violet line. If, again, we take a salt of strontium, which was one of the ingredients in red fire, it colours the flame crimson, and by the eye the flame can scarcely be distinguished from the colour of the lithium flame; but in the spectroscope there is no possibility of doubt, the spectrum of strontium consists of a group of several lines in the red and orange, and a fine line in the blue end of the spectrum.

If a higher temperature than that of the Bunsen flame is required, the blow-pipe (Fig. 24), may be resorted to; in this the quantity of air and coal-gas is varied at pleasure, and a very high temperature may be obtained.

We might proceed thus to examine all the elementary substances one by one, but to observe the spectra of the metals, it will be found necessary to use a higher temperature still, and for this purpose the electric arc or spark is employed. If a temperature only slightly greater than that of the blowpipe flame is used, the spark from an induction coil worked by five Grove cells may be taken, as shown in Fig. 25, the Leyden jar not being employed; a few metallic lines will then be seen, and a background consisting generally of bands of light here and there.

If a higher temperature still is used, the jar may be thrown into the circuit, upon which the spark will become more intense, according to the power of the coil and size of the jar; or the electric arc may be employed (Fig. 26). The spectra thus obtained are much more complex, as the vapours are much more dense; the spectrum of iron, for instance, when thus examined is found to consist of no less than 460 lines, many of which are situated in the green part of the spectrum.

With regard to solid and vaporous bodies, the electric lamp affords a very handy method, when

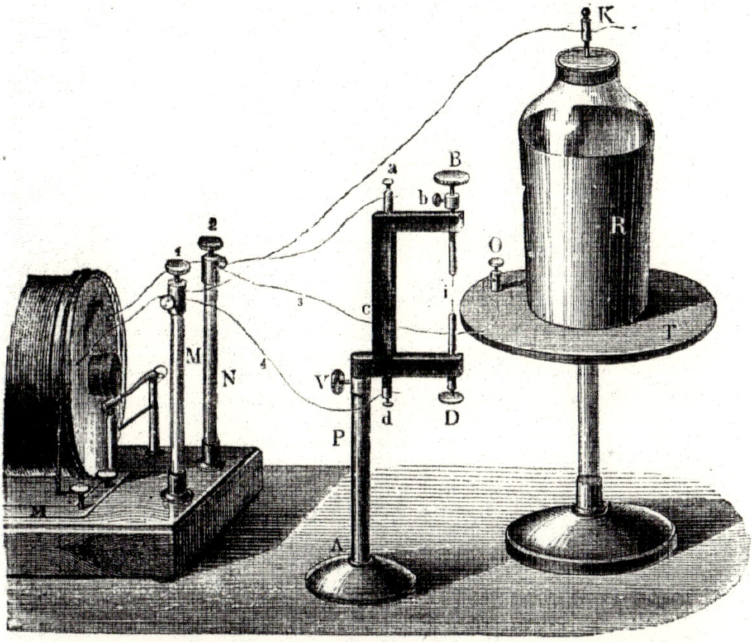

FIG. 25.—Arrangement for determining the spectra of metals by means of the electric spark, showing Induction Coil, Leyden jar, and spark stand.

properly employed, of examining and exhibiting the spectra of these bodies to large audiences.

But there are a great many gases which the spectroscopist also has to study, and to study with the greatest care; and here, I am sorry to say, the electric lamp utterly fails us. The light which we get from a gas by the electric discharge is so feeble that it is quite impossible to throw its spectrum on the screen, so as to be observed by large audiences, for we cannot render strontium incandescent in the way in which

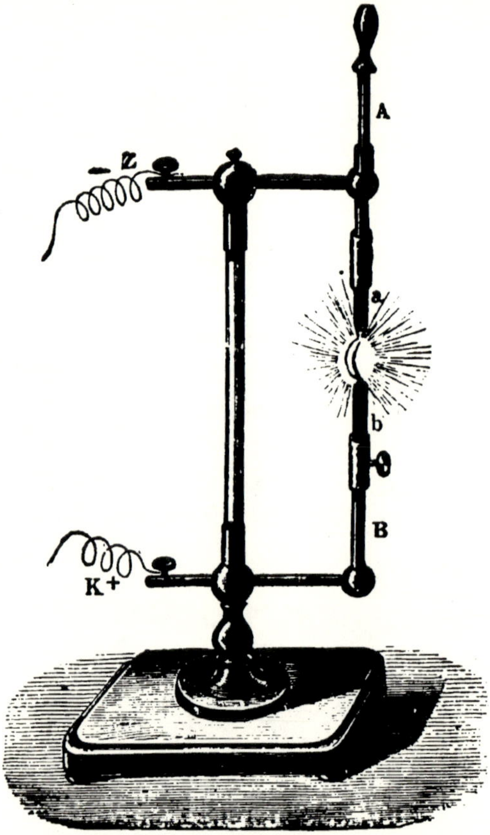

FIG. 26.—The Electric arc.

we render incandescent sodium and the other substances I have brought before you. But we have other means of examining the spectra. I have here some tubes containing hydrogen and other gases at different pressures (Fig. 27), and when we wish to study the spectrum of a gas, we do it in this way: we enclose it in a tube, and send a current through it by means of an induction-coil. If we pass a stream of electric sparks through a tube containing hydrogen at the pressure of one atmosphere, we shall see that the colour of the incandescent gas is a bright carmine red, the spectrum of which can easily be observed by placing the spark tube in front of the slit of one of the spectroscopes before described. This arrangement is one that is in daily use in many of our laboratories, and it must be borne in mind as being the *modus operandi* by which a great deal of the work has been done to which I shall have to allude shortly. If again we take a tube which contains hydrogen which has been extremely rarefied, and pass a series of electric sparks through it, instead of having the brilliant red colour, we shall have a pale greenish spark, quite different from the former. This great difference is due to the difference in the pressures of the hydrogen of the two cases.

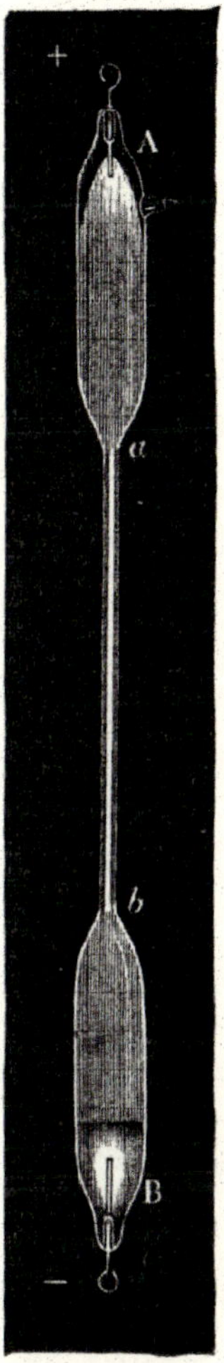

FIG. 27.—Geissler's tube, showing electric discharge.

The two spectra are equally distinct, the red light shows three splendid lines, one in the red, another in the bluish green, and the third in the violet, together with a considerable amount of continuous spectrum, whilst almost the only spectrum which can be obtained in the second case, is a single green line in the same position as the former green line spoken of. There is also this difference which will be observed, that the green line obtained from the tube at the atmospheric pressure is very broad and indistinct at the edges; and that the line as seen in the almost vacuous tube is very thin, comparatively speaking, and perfectly sharp and well defined. If we were to take another tube, with a pressure somewhere between the two already mentioned, it would be seen that this green line was not so wide and woolly as in the tube at one atmosphere, and yet not so sharp and well defined as in the almost vacuous tube. Thus it will be seen that this widening out of the line is due to the difference of pressure.

Spectrum analysis, then, teaches us this great fact, that solids and liquids give out continuous spectra, and that vapours and gases give out discontinuous spectra; that is to say, that we get bright lines in different parts of the spectrum, instead of having an unbroken light all over the spectrum. I might vary this statement by stating broadly that the radiation or giving out of light by solids and liquids is a general one, and that the radiation or giving out of light by gases and vapours, instead of being general, is in the main a selective one.

The tubes, to which reference has already been made, put us then in complete possession of a point which has already been arrived at by two different lines of investigation. A few years ago, Dr. Frankland, in investigating the spectrum of hydrogen, which, as you know, according to the statement I have just made, ought to give a discontinuous spectrum, discovered that when observing the spectrum under very great pressure, he got a white light and a continuous spectrum. Afterwards, Dr. Andrews, another Fellow of the Royal Society, who was working at the theory of vapours and the theory of liquids from a perfectly different standpoint, and who never thought of using a spectroscope at all, arrived at the conclusion that it was quite possible that vapours might be so condensed in almost every case, that by crushing them together, so to speak, you might really arrive at a liquid form of the vapour which you might choose to investigate. I hope you will not think that these high physical investigations are not practical enough. Let me remind you that we do not know what they may lead to.

Not only did Dr. Frankland determine that very dense gases and very dense vapours gave continuous spectra, but in another research, in which I have had the honour of being associated with him, we have shown that the spectrum of a vapour or of a gas does very much more than tell us merely what the gas or vapour experimented upon is; it in fact tells us something of the physical condition of that gas or vapour, that is to say, whether it is very rare or whether it is very closely packed together—whether it exists under a low or a high pressure. Very fortunately for us, this is an investigation which has not only an immense application in every chemical experiment with which the spectroscope has to do, but it has its story to tell and its aid to give concerning every star that shines in the heavens. We may state generally that, beginning with any one element in its most rarefied condition, and then following its spectrum as the molecules come nearer together, so as at last to reach the solid form, we shall find that spectrum become more complicated as this approach takes place, until at last a vivid continuous spectrum is reached.

Spectrum analysis, then, if it merely differentiated between gases, vapours, solids, and liquids, and between gases and vapours in different states of pressure, would really be a new chemistry altogether; and I have no doubt that the time is not very far distant when, not only in the chemist's laboratory, but in a great many applications of the physical sciences, the spectroscope will be considered as necessary, and will be almost as much used, as a chemical balance, and the sooner that time comes the better.

But not only are we able to differentiate between different bodies, but the most minute quantities of substances can be determined by this method of research. The thing seems so impossible, that you may, some of you, feel inclined to doubt my assertion when I tell you, for instance, that Kirchhoff and Bunsen have calculated that the 18-millionth part of a grain can be determined by the spectroscope in the case of sodium; that is to say, if in anything which I choose to examine by means of my spectroscope the quantity of sodium present amounts only to the 18-millionth of a grain, the spectroscope is perfectly competent to take up that minute quantity, and bring it out into daylight, so as to be detected with certainty. This reaction of sodium is so delicate, that if we examine any flame burning in air, we almost invariably find sodium in it, for every particle of dust is impregnated with a sodium salt, probably sodic chloride. This is not to be wondered at, as two-thirds of the earth's surface is covered by sea, which contains a considerable amount of sodium salts, and the fine spray, which is continually caused by the dashing of the waves, evaporates and leaves *minute* specks of salt which are carried over the whole land, and make themselves visible in our spectroscopes. Take another instance. Lithium is a substance the knowledge of the existence of which as a common element we owe entirely to the spectroscope ; the 6-millionth part of a grain of this can be detected. If we examine anything for lithium, and do not get the characteristic red line, we know that not even the 6-millionth of a grain is present. Strontium, again, can be discovered if only a millionth part of a grain is present. So much for the great power of spectrum analysis in its physical applications, and its dealing with minute quantities of the elements which we know already, and this of itself would be of enormous importance.

But the spectroscope does not stop here; it discovers the known elements under conditions where detection seemed almost impossible, and in which the old chemistry was powerless to help us. Let us take again, for instance, lithium. Lithium was only known formerly to exist in four minerals; it is now known, thanks to the spectroscope, to exist almost everywhere. If we were to take the ash of a cigar and introduce it into a colourless gas flame and examine the coloration with the spectroscope, we should get a spectrum of lithium; and if we analysed in the same way the ash of milk, or the ash of blood, or of grapes, tea, sugar, &c., we should also find it. Dr. Miller has shown that, in the Wheal Clifford Mine, 800 lb. of this salt are given every twenty-four hours, though before the advent of spectrum analysis no lithium was known to exist there. It has also been found in meteoric stones, in the water of the Atlantic, &c. Surely this is an application of very great importance.

Another extremely important point about the spectroscopic analysis is, that although we may have to analyse a complicated mixture of substances, the spectroscope is perfectly competent to deal with them. The characteristic lines for each element must stand out and be visible whether the substance be simple or complex. Thus, for instance, if we mix together some sodium and lithium, and place some of the mixture in a flame, we shall see nothing but the brilliant yellow colour due to sodium, the crimson flame of the lithium being entirely hidden. A moment's examination with the spectroscope, however, is sufficient to show us that both lithium and sodium are, without the slightest doubt, present in the flame; for both the yellow and red lines stand out as distinctly as they did when the simple salts were experimented with. The presence of lithium, indeed, may be detected, even if it be mixed with ten thousand times its bulk of sodium compounds.

But, further, spectrum analysis is not satisfied with showing us sources of known elements. It discovers new elements altogether. In 1860, Bunsen happened to be examining with a spectroscope the result of one of his analyses of the waters of a spring near Dürkheim, and he saw some lines which he had never seen before, although he had very carefully mapped the spectra of the known elements. Bunsen, as you know, is a very resolute chemist, and what he did was this. Having faith in his instrument, he evaporated no less than forty-four tons of the water of this spring, and out of these forty-four tons he got about two hundred grains of what turned out to be a new metal, which he called Cæsium. Rubidium was the next metal which was discovered in this way. Take another instance, the discovery of thallium by Mr. Crookes. Mr. Crookes was working with a seleniferous deposit from the Hartz mountains, when, by the aid of the spectroscope, he discovered this metal, which, I am informed, is now extensively used in the manufacture of fireworks. The spectrum of this metal is extremely distinct and beautiful, and you will understand why it has been named thallium, from the Greek word for a twig, on account of the beautiful green colour of the single line ordinarily visible.

A fourth element has been discovered by means of the spectroscope by two German chemists, Professors Reich and Richter, who were experimenting on zinc blend, and found two unknown indigo bands in the spectrum, which they successfully traced to the existence of a small quantity of a new metallic element, which has been named Indium.

You all know how important chemical analysis is in thousands of things connected with the arts, manufactures, and commerce, in detecting adulteration for instance, and in these matters the spectroscope gives our chemists a power which was undreamt of a few years ago.

There is another very beautiful application of the spectroscope which perhaps many of you will say is of more practical importance than those I have already brought to your notice. You know that, in the Bessemer process five tons of cast iron are turned into cast steel in twenty minutes. Now steel is only cast iron minus some carbon. It is clear, therefore, that the process depends upon getting rid of the carbon. How then can the spectroscope aid us in determining the time at which the carbon is got rid of? Nothing is more easy. The heat from the incandescent iron is so intense that the vapour of the different substances mixed with it is visible above the retort in which the metal is placed, and we get, so to speak, an atmosphere of incandescent vapour surrounding the cast iron. When we examine these incandescent vapours by means of a spectroscope, it is found that the spectrum changes very considerably at different times during the combustion of this cast iron. Now it so happens, that the process of conversion is such a delicate one that a mistake of ten seconds either way spoils the whole five tons which are being operated upon. You will see in a moment, therefore, that this is a case in which any rule-of-thumb or very rough method might now and then lead to a mistake; but when the spectrum of these incandescent vapours thrown out by the cast iron is examined very carefully by means of a spectroscope, it is found that at the first the spectrum of carbon is quite visible, but at the right moment, which has been found by practice, that spectrum disappears, the combustion having been sufficient. All we have to do now, to ensure the charge being properly turned out, is, therefore, by means of the spectroscope, simply to watch certain lines in the spectrum, and when they show signs of disappearance say "Now," and the thing is done without any possibility of error. This is an instance of the practical application of the spectroscope in one direction; now let me give you one in another.

When Dr. Bence Jones wished to determine some questions connected with chemical circulation, he employed the spectroscope with great success. Many of you, at the first blush, would be inclined to say it was not very likely that the spectroscope would help us here. If it were a question, for instance, of our own chemical circulation, we would not relish the idea of being converted into an incandescent vapour for the pleasure of testing a theory. But, fortunately, there are such things as guinea-pigs, and Dr. Bence Jones, by studying the vapours of the ashes of these animals, has arrived at some results of extreme importance. His *modus operandi* was as follows:—He gave the guinea-pigs chloride of lithium, and then the question was to find, by burning the ashes of the different parts of the guinea-pigs, variously removed from the fountain of circulation and from the ordinary ducts of the body, to ascertain how long it took lithium to get absorbed into every part of the body. The most distant part, as far as circulation goes, is the lens of the eye If, then, we give a guinea-pig chloride of lithium, then kill the guinea-pig, and examine the ash of the eye lens, say three hours after the lithium has been taken into the system, and if we find the lithium line in the spectrum of the ash vapour where no lithium was before, that is to say, if by means of the spectroscope we see that line which we have seen characterizes the lithium spectrum, we know that the chemical circulation of the body is such as to take lithium through the body to that particular point of the circulation in that time. In the human subject Dr. Bence Jones has hit upon a very practical method of arriving at something like the same conclusion, by examining the spectra of the ashes of cataracts.

So far as I have dealt with the applications of the spectroscope, up to the present time, I have dealt in the main with the application to chemistry and to physics—in other words, to the examination of light given out by terrestrial substances; but I must now, with your permission, take you to the skies—reminding you that at present I am merely dealing with the giving out of light—and deal with light emitted by celestial objects. We shall afterwards have to deal with the stopping or absorption of light, by vapours and other transparent media when the light passes through them.

I have already referred to the special fittings that were necessary for the application of the spectroscope to the telescope, and I think on carefully looking at the engraving (Fig. 17) representing a star spectroscope, you will see exactly how the spectroscope is applied to a telescope. We must now go a little more into details. One class of spectroscopes, as applied to telescopes, is arranged for observing the spectra of the stars, nebulæ, &c., and another with a much greater dispersive power for observing the spectrum of the sun. In both spectroscopes the

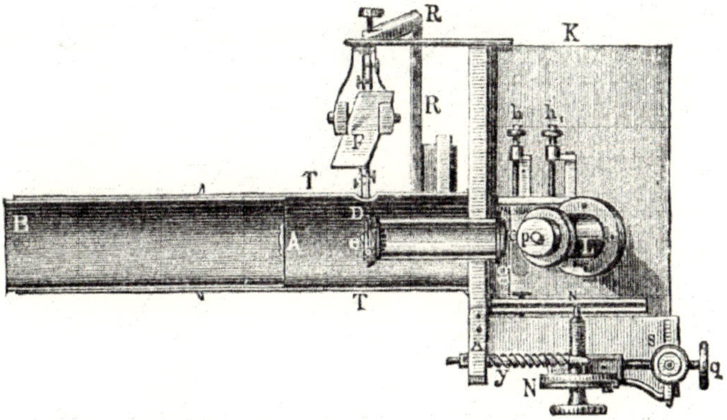

FIG. 28.—Side view of Star Spectroscope, showing the arrangement by which the light from a spark is thrown into the instrument by means of the reflecting prism *e*, by a mirror, F.

arrangements employed are similar, and resemble those of the instruments that have been already described. A finder on the side of the large telescope enables the image of the star to be brought on the slit, while, in the case of the sun, its image is received on the slit screen, and any part of the image may be brought on the slit by mere inspection. The spectroscope is attached to the eye-piece end of the instrument, and the image formed by the telescope is received on the slit plate. Arrangements are necessary in the case of the star spectroscope for widening out the spectrum: this is done by a cylindrical lens, as before explained; and for obtaining a spectrum of comparison, this is done by reflecting into the instrument the light emitted by an electric spark.

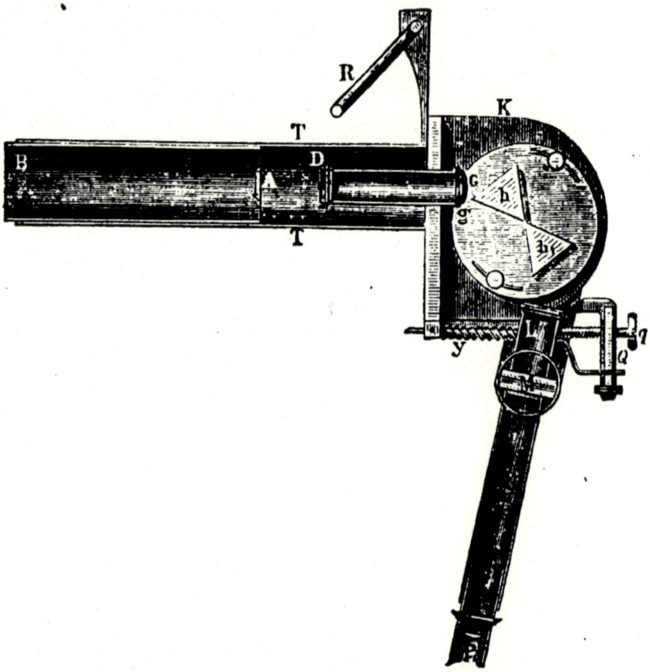

FIG. 29.—Plan of Star Spectroscope. T, Eye-piece end of telescope; B, Interior tube, carrying A, cylindrical lens; D, Slit of spectroscope; G, Collimating lens; *i h*, Prisms; Q, Micrometer.

In the star spectroscopes, the number of prisms, and the consequent deviation and dispersion, is small. The accompanying woodcuts will make their detailed construction quite clear. In the case of sun spectroscopes, the deviation and dispersion required are large, the deviation amounting to over 300°; that is to say, the ray of light is bent through almost a complete circle; the light from stars is dim, and many prisms cannot be employed to widen out the spectrum, but, in the case of the sun, there is light

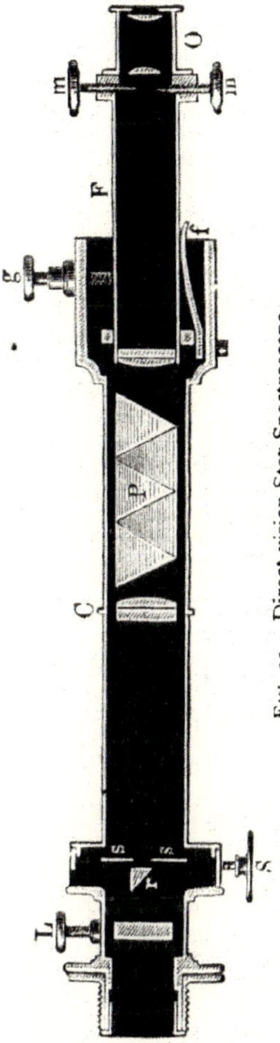

FIG. 30.—Direct-vision Star Spectroscope.

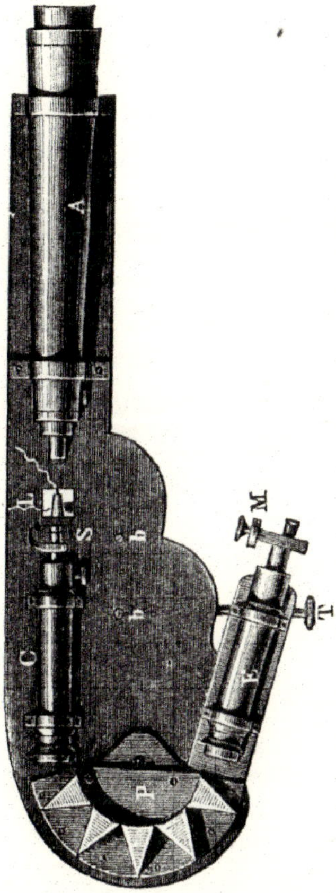

FIG. 31.—Sun Spectroscope. A, Telescope; S, Slit; P, Prism plate; E, Observing telescope; M, Micrometer.

sufficient to give us a bright spectrum after it has been enormously dispersed.

Figs. 32 and 33 show very powerful spectroscopes, to be attached to the telescope for observing the spectrum of the sun. One peculiarity of the instrument in Fig. 33 is that the ray of light having passed once through the lower part of the train of prisms, is received by a right-angled prism, which totally reflects the light twice, sending the ray of light back through the upper part of the same prisms, when it is again refracted; we thus have, by using these prisms, the same effect as if thirteen prisms had been employed. The ray of light enters the instrument by the lower tube, and after

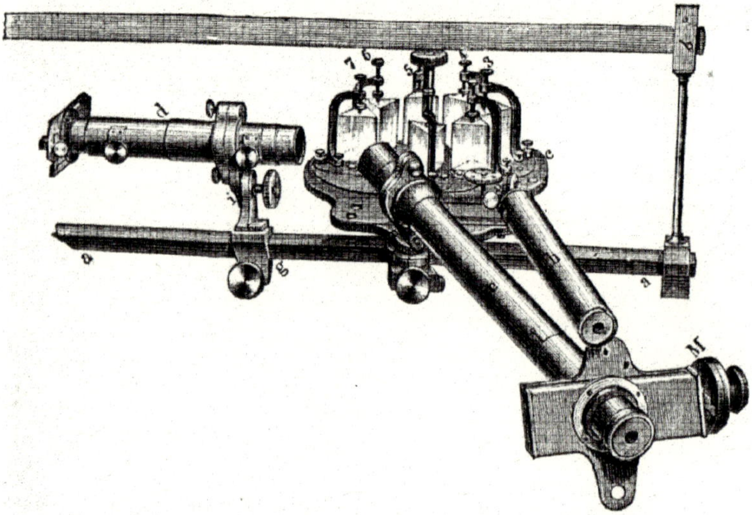

FIG. 32.—Sun Spectroscope. *d*, Collimator; *e*, Observing telescope; *h* and M, Two micrometers; 1, 2, 3, 4, 5, 6, 7, Prisms.

passing first through the lower half of the prisms, and back through the upper half, is received in the upper tube, and reflected upwards for convenience of observation. These prisms are so arranged, that whatever part of the spectrum is being observed, they are always at the angle of minimum deviation for this part of the spectrum, a very important point, as if this is not attended to the spectrum loses much of its brilliancy and sharpness. This is done either by attaching the prisms to a spring of ebonite or gun metal moving on a fixed point near the first prism of

FIG. 33.—Sun Spectroscope arranged for photography.

the series, as in the arrangement shown, or each prism may be attached to a radial bar acting on a central pin, as shown in Fig 34.

In the first place, then, what does the spectroscope tell us with regard to the radiation from the sun and the stars? And here I ask you to neglect and banish from your minds for a time any idea of those dark lines in the solar spectrum that I drew your attention to on a former occasion. I hope I shall be able to explain them satisfactorily to you afterwards, but for the present I wish you merely to take the fact that our sun, but for the dark lines, would give us a continuous

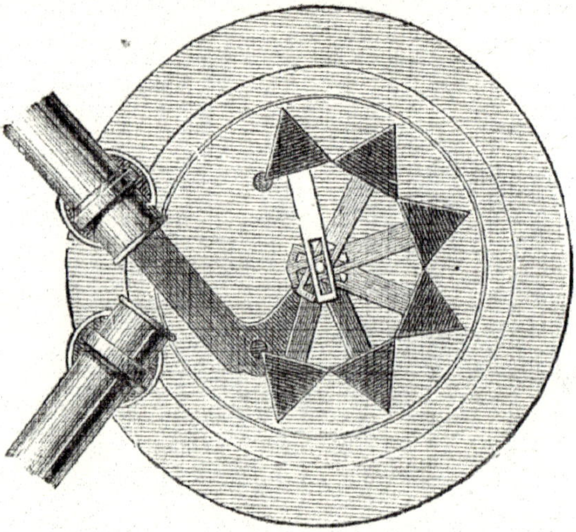

FIG. 34.—Automatic arrangement for securing the minimum deviation of the observed ray.

spectrum. The spectrum of the stars is very much like the spectrum of the sun. In Fig. 35 is seen a representation of the spectra of two stars, *a* Orionis and Aldebaran, mapped with the minutest care by Dr. Miller and Mr. Huggins. In both cases we should have a continuous spectrum but for the presence of the dark lines. I think you will see in a moment what I am driving at. Suppose the sun or stars composed of only sodium vapour, for instance, it is

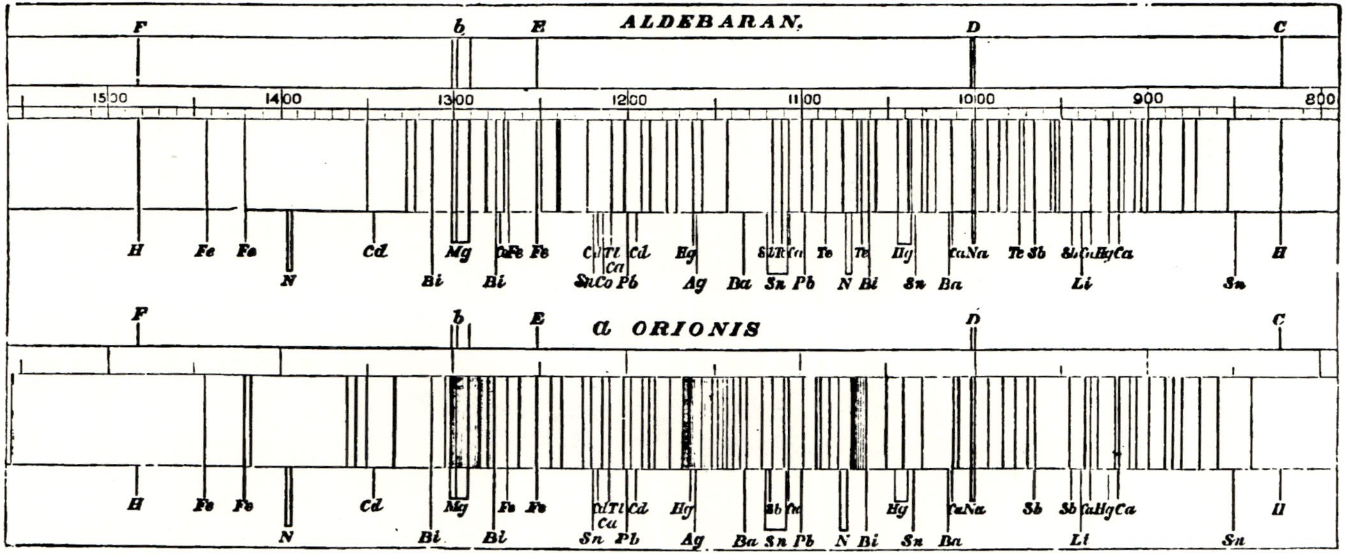

FIG. 35.—Spectra of Aldebaran and α Orionis. (Huggins and Miller.)

clear that their light analysed by the prism would give us no great indication of a continuous spectrum, we should merely get one bright line in the orange. But neglect the dark lines for a moment: dealing merely with the continuous spectrum of the sun and star, it shows that we have a something, whether it be solid or liquid, or whether it be a dense gas or vapour, competent to give us a continuous spectrum. So we are justified in assuming that sunlight and starlight proceed from the incandescence of a solid, liquid, or dense gas or vapour. Again, suppose that instead of looking at the sun or the stars we observe the moon, as Fraunhofer did, as has been before stated, what will happen? We get a second edition of sunlight, in exactly the same way as we should get a second edition of sunlight in the case of a reflection of it from a mirror; and therefore, if proof of such a thing were needed, the spectroscope is perfectly competent to show us that the moon gives us sunlight second-hand. The same in the main with Jupiter, Venus, Mars, and the other planets. If we study them and observe the dark lines we find that the lines which we observe are generally the same as those which we find in the spectrum of the sun. There are other points to which I shall have to draw your attention on a future occasion, but, on the whole, the teaching of the spectroscope is that all those planets are lit up by sunlight, as we know them to be.

But we have not yet exhausted the wonders of the celestial field; we have dealt merely with the sun and moon, the stars and planets. What about the nebulæ, those strange, weird things, dimly shining in the depths of space, both to the eye and in the telescope obviously and distinctly different from anything in the shape of the sun or stars? The appearance of these peculiar bodies is sufficient to show us that we have here something very different from the sun or moon. What is it? You all know as well as I do that ever since nebulæ were discovered mankind have wondered at them, and wanted to know what they were; and you are also aware that it was not settled and could not be settled before the advent of the spectroscope, but that it could be settled in five minutes after that event. Mr.

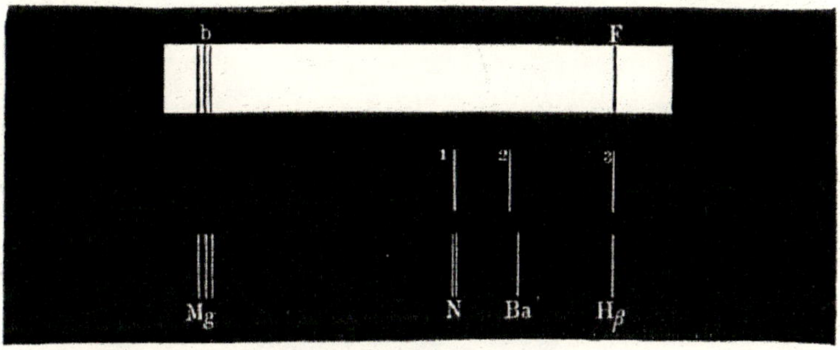

FIG. 36.—Spectrum of the nebulæ.—1, 2, 3, lines observed. Above, the solar spectrum is shown from *b* to F; below, the bright lines of magnesium, nitrogen, barium, and hydrogen, in the corresponding part of the spectrum.

Huggins, who first observed the spectrum of a nebula, found that, instead of the continuous spectrum with which you are familiar in the case of the sun and the stars—always asking you to neglect the Fraunhofer lines, which I shall explain afterwards—the light which he got from the nebula consisted merely of three lines. He was exceedingly astonished, so much so that he thought the instrument might be out of order. However, it became perfectly clear to him in a very short time that there was no mistake at all, and that all that the light which came from the nebula could do was to give him these three faint lines. No doubt you have anticipated my explanation. The nebulæ are composed of tenuous gases or vapours. After what I have said about the way in which the spectroscope at once picks out the difference between a solid or liquid, and a vaporous or a gaseous body, you will see at once that these three bright lines indicate that the nebulæ, instead of being composed of solid, liquid, or densely gaseous bodies—instead of being like the sun or stars—are really composed of rare gases or vapours. Mr. Huggins was enabled, in fact, to determine the gas in one instance, for one of the

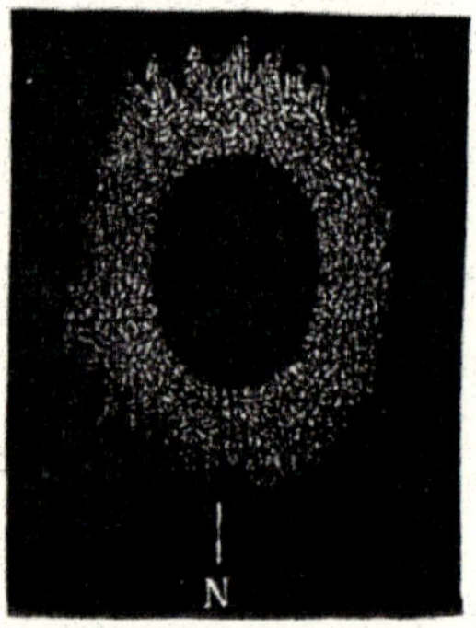

FIG. 37.—Ring nebula in Lyra, with its spectrum.

lines he found was coincident with one of the principal lines in the spectrum of hydrogen, one of the other lines possibly being due to nitrogen. And now comes another extremely important point, showing the importance of studying the most minute changes in gaseous spectra, for Mr. Huggins, who knew the spectrum of hydrogen and the spectrum of nitrogen well, and who knew how extremely complicated those spectra are at times, was much astonished at finding only one line of hydrogen and one of nitrogen, and attempted to account for the singleness of the lines, first, by assuming a condition of the gas different from anything we meet with in our laboratories, and again by assuming an absorbing medium in space. But after Dr. Frankland and myself had made some observations on the spectra of hydrogen and nitrogen, we found it was perfectly easy to obtain, and sometimes when one did not want it, a spectrum

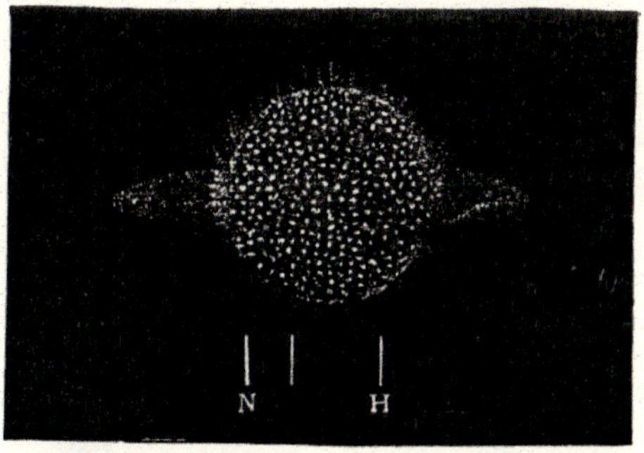

FIG. 38.—Planetary nebula in Aquarius, with its spectrum.

of hydrogen or of nitrogen giving only one line, or nearly so; so that by comparing the conditions which were necessary to obtain these conditions in our tubes with the conditions of the nebulæ, it was quite possible to make at all events a rough guess at what is the constitution of the nebulæ, as far as pressure or molecular separation goes. We find, for instance, this single line of hydrogen, and a nearly single line of nitrogen, when the pressure is so slight that you would say that the tube really contained nothing at all, and when, moreover, the temperature is comparatively low. Now, not only is this a fact which we are quite prepared to assert merely on the evidence rendered us by these tubes, but I think you will acknowledge that it is entirely in accordance with everything we know astronomically on this subject.

For the next application of the spectroscope in this direction, let us take a comet. The appearance

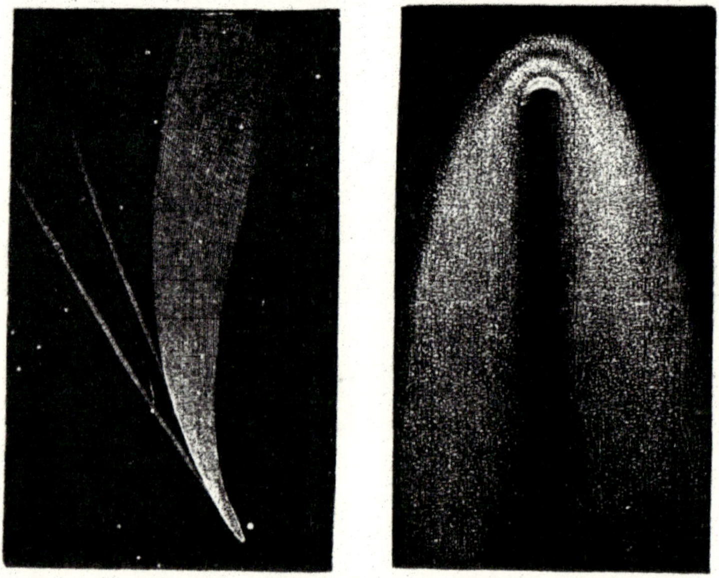

General view. Head and envelopes.

FIG. 39.—Views of Donati's comet.

of a comet is probably well known to many, who will recollect the form of Donati's comet. Although, as you know, that comet appeared only about ten years ago, unfortunately it came too early for us to learn anything about it by means of the spectroscope. We have, first of all, an extremely bright nucleus; then a kind of semilune of greater brilliancy than the rest of the head; then what is called the coma, and the tail. The question which the spectroscope had to put to the comet was—of what is the nucleus composed, and of what is the tail composed. Professor Donati, and Mr. Huggins especially, to whom we owe so much for his work in this direction, has made some observations on two small comets—I am sorry they were not

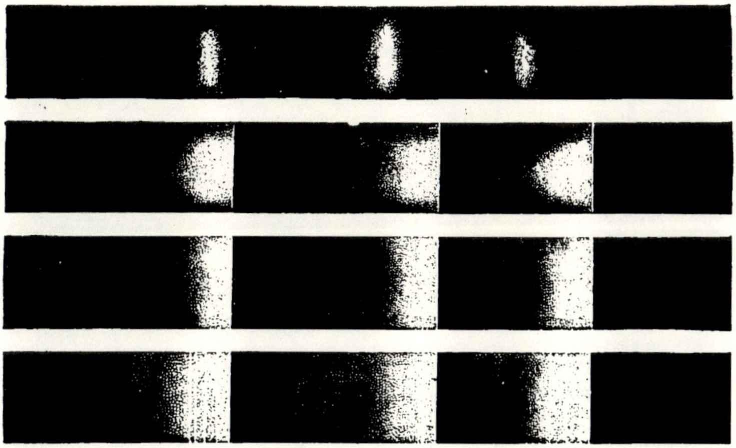

FIG 40.—1. Spectrum of Brorsen's comet; 2, Spectrum of Winnecke's comet; 3, Spectrum of carbon in olefiant gas; 4, Spectrum of carbon in olive oil. (Huggins.)

larger—with considerable success. He found that in the comets he examined, the head gave out a light which very strongly indeed resembled the spectrum of carbon vapour. The spectrum of carbon taken with the spark in olive oil and in olefiant gas differs slightly: the spectrum as obtained from the latter consists of three bands or waves of light, which commence tolerably bright and sharply on the red side, and become gradually fainter towards the more refrangible side. These bands are severally situated in the beginning of the green, in the true green, and in the blue portions of the spectrum. Mr. Huggins has also observed the spectrum of Encke's comet, and has confirmed the result that he previously obtained, viz. that the spectrum of the comet is identical with the spectrum of carbon, as taken in a hydrocarbon. I should like to draw your attention, if there were time, to the way in which these spectra of the carbon spark taken in oil and in olefiant gas, differ.

I have not yet completed all I have to say on the subject of radiation. If, as we have already seen, we take a tube containing incandescent hydrogen and pass a series of intense electric sparks through it, we see that it gives out a red light, which may remind you of some other specimens of radiation which is supplied us by the skies. I allude to the red prominences which are seen around the sun, not in ordinary times, but when the sun is eclipsed. This representation gives you a good idea of what really is seen when the sun is eclipsed. In Fig. 41 we have as it were a black sun instead of a bright one, which is really nothing but the body of the moon. Around this we have a ring of light, which is called the corona; and here and there in this corona we have what are called red flames or red prominences. These red prominences have also, on closer observation, been found to be only local aggregations or heapings up of a red layer which envelopes the sun. Here then, it was quite possible that if the newly invented spectroscope were set to question these things, we should see at once

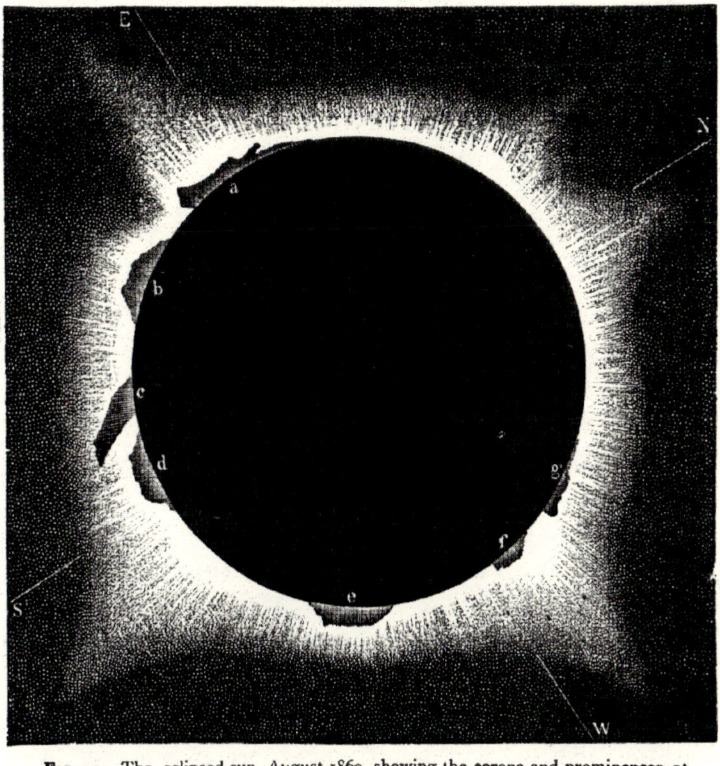

FIG. 41.—The eclipsed sun, August 1869, showing the corona and prominences at *a, b, c, d, e, f, g*.

whether they were solid or liquid, or whether they were gaseous or vaporous. If we got a continuous spectrum from these red things, we should know that they were solid, or liquid, or densely gaseous. If, on the contrary, we got a bright line spectrum, we should know we were dealing with a gas or vapour. You also see that, as the light is red, the chances were that they were not solid or liquid; and then you further see that if the things do consist of a light which does give us lines, a determination of the exact position of the lines, and a comparison of these positions with those of hydrogen, sodium, magnesium, barium, or anything else, would teach us what these things were.

Another point was also very obvious to those who are familiar with these inquiries, namely, that if these prominences really consisted of gas, by the use of a powerful spectroscope it was perfectly unnecessary to wait for eclipses at all. The reason for this will be clear on a little consideration. If we take a continuous or unbroken spectrum and apply successively a number of prisms, the spectrum will become proportionately lengthened, and therefore more and more feeble, and in fact we can thus reduce the light to any degree required. If now, on the other hand, we take a spectrum which consists only of bright lines, say of one line in the red and another in the blue, and as before apply successively a number of prisms, we shall, it is true, increase the length of the spectrum, that is the distance between the two lines, but this will be all; the additional prisms have no power to alter the width of the lines themselves, for we have seen that these are simply the images of the slit; their light, therefore, will only be slightly enfeebled owing to reflection merely. Thus if we have a mixed light to analyse, part of which comes from a source giving out a continuous spectrum, and the rest that of a glowing gas, although when working with a single prism no lines may be visible on account of the brightness of the continuous spectrum, yet by using say five or seven prisms we can so dilute the continuous spectrum as to render the bright lines of the glowing gas clearly visible. The case of the red flames round the sun is a case in point. They are invisible to the naked eye and in telescopes on account of the intensely illuminated atmosphere, which also prevents anything like bright lines being observed from these red flames until the bright continuous spectrum has been much reduced. When this has been done, the bright lines of the spectrum, should there be any, will appear on a comparatively dark background. Now M. Janssen (who was sent out by the French Government to observe the eclipse which was visible in India in 1868), Major Tennant, and others, had no difficulty in recognizing in a moment, when the sun was eclipsed, that these things really did consist of gases or vapours; and M. Janssen, a very careful observer, had no difficulty in determining that the gas in question was really hydrogen gas. M. Janssen and myself were also enabled to determine this by observations *on the uneclipsed sun*, by means of the new method of observation I have just sketched out. The accompanying woodcut (Fig. 42) shows the spectrum which is observed from these solar prominences. The spectrum of the prominences is shown in the upper, and that of the sun in the lower half of the engraving. This method is very easy to understand if you bear in mind the engraving of the large spectroscope for solar work, and recollect that when we wish to examine the regions round the sun, the light of the sun is allowed to fall on the slit in such a way that one half of the slit at the focus of the object-glass of the large telescope is occupied by the brilliant image of the sun, and the other half is fishing, so to speak, around the limb or edge of the sun, so that if there is anything at all around the limb, the spectroscope, in the —to the eye—unoccupied part outside the image, picks up this something, and gives us its light sorted out into its proper bright lines in the spectrum. This spectrum shows that there is first a bright line (Fig. 43), in the red, marked C, which is absolutely coincident with

FIG. 42.—Spectrum of the sun's photosphere (below) and chromosphere (above)

a prominent dark line in the solar spectrum. Now this is a black line which, by repeated observations, we know corresponds in degree of refrangibility exactly with one of the lines given out by glowing hydrogen when examined in one of these tubes with the electric spark. When, therefore, we get any substance around the sun reporting its light to us, it is perfectly obvious, I think, that if the bright line really be coincident with this dark line, that something is

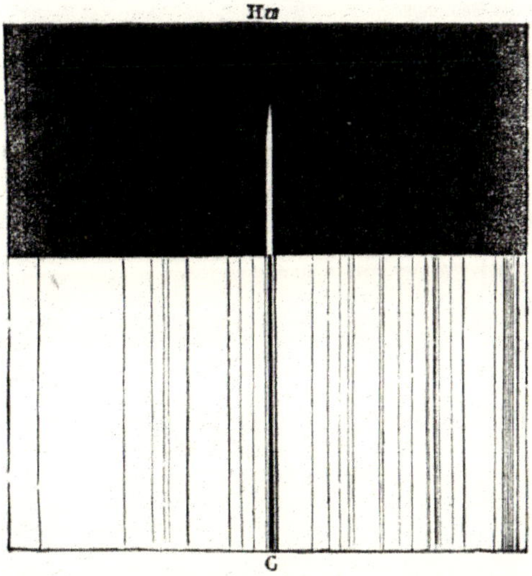

FIG. 43.—C line bright in chromosphere, dark on sun.

probably hydrogen. This was one of the first lines determined by M. Janssen in the eclipse of 1868. There is another bright line absolutely coincident with a dark line known to correspond in refrangibility with another line given out by hydrogen in the green part of the spectrum, marked F in the figure. This, then, is further proof in favour of hydrogen. Now notice a great difference between the shape of this line and the red line which I drew your attention to just now. An enlarged representation of this line is shown in Fig. 44.

You will bear in mind what I told you about the effect of pressure in altering the spectrum of hydrogen, and that one of the most obvious effects of increased pressure was to increase the thickness of what is called the F line—the line now under consideration. You will see here that the widening

FIG. 44.—F line in chromosphere, showing widening near the sun.

of the F line, the green line of hydrogen, really indicates a thickening due to pressure. In that way we have been able to determine approximately the pressure of these circumsolar regions which the spectroscope has determined to be occupied by an envelope of hydrogen gas, mingled sometimes with other vapours, which envelope I have termed the Chromosphere. When the pressure of the chromosphere is completely determined, we shall be probably enabled to determine the temperature of the sun.

A line in the violet, again, corresponds with a dark line in the solar spectrum, which is coincident with a third line of glowing hydrogen which we have before spoken about, and there is still another coincident line. A line in the yellow of the spectrum will also be noticed. This is one which has caused a great deal of discussion, for it is not coincident with any line of any known terrestrial substance. A number of short lines are also shown in the engraving, which will be seen to correspond to the part of the chromosphere which is denser, for then the F line of hydrogen has become broad where these lines are seen : these lines show that in the layers of the chromosphere nearest to the sun a number of other substances exist, amongst which may be mentioned magnesium, iron, and sodium. The reason that these do not reach up so far from the body of the sun is that their vapours are very much heavier than the gas hydrogen, which is the lightest terrestrial substance known.

Such are a few of the practical applications of the spectroscope as applied to the radiation of light. There are other classes of facts relating to the absorption of light, and these will form the subject of my next lecture.

LECTURE III.

On the last occasion, the subject which we dealt with was the radiation or giving out of light by bodies in different states—that is to say, by solid or liquid bodies, or gaseous or vaporous ones. We have now to deal with the action of the prism upon light under some new conditions—conditions which I purposely withheld from you in the last lecture. Light is not only given out, or *radiated*, but it may be stopped or *absorbed* in its passage from the light-source to our eye, if we interpose in the path of the beam certain more or less perfectly transparent substances, be they solids, liquids, gases, or vapours. I will recall one or two of the experiments which were described on the former occasion, in order that you may see exactly how the perfectly distinct classes of phenomena due to radiation and absorption really run together. You will recollect that I pointed out to you that radiation, or the giving out of light, might be continuous or might be selective, and I am anxious now to show you that radiation is exactly equalled by absorption in this matter; that absorption may also be continuous or selective. We have before taken as an instance of continuous radiation a continuous spectrum obtained by using the electric lamp or a lime-light; that is to say, an example of the general radiation which you get from an incandescent solid—the carbon points of which the poles of the lamp are composed, or the solid lime. You will remember that if we take the spectrum of a vapour—as, for instance, that of strontium or thallium—we find that the continuous spectrum is altogether changed, and that, in the place of that beautiful rainbow band, continuous from the red end of the spectrum to the violet, we really only get lines here and there, which are due to the selective radiation, and opposed to the general radiation which we spoke of in the continuous spectrum just now. I might have chosen other substances besides strontium and thallium, but I mentioned the spectra of these substances when we were considering the question of radiation. What I have to dwell on now is, that the absorption or sifting of light by different bodies is very like radiation in its results—that is to say, in some cases we have an absorption which deals equally with every part of the spectrum, and in other cases we have absorption which only picks out a particular part of the spectrum here and there to act upon. But there is one important point to be borne in mind; when dealing with absorption we must always have a continuous spectrum to act upon. If we had a discontinuous spectrum to act upon, the thing would not be at all so clear. Having this continuous spectrum, the problem is, what the action of the different substances on the light will be. Let me give you an instance of general absorption. If we take the continuous spectrum above referred to, and interpose a piece of smoked glass, or, better, a piece of neutral-tinted glass, you will find that the substance will cut off the light and deaden the spectrum, so to speak, throughout its whole length. This neutral-tinted glass, then, has the faculty evidently of keeping back the light, red, yellow, blue, green, violet, and so on; and is an instance of general absorption. A very dense vapour would furnish us with another similar instance.

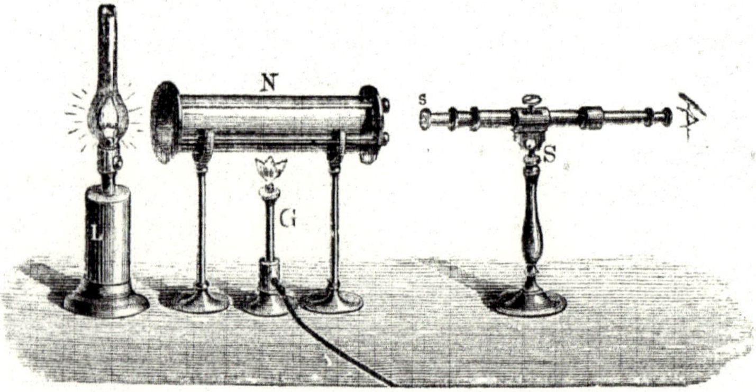

FIG. 45.—Method of observing the absorption of a vapour.

Now, instead of using the neutral-tinted glass, we will introduce a piece of coloured glass, the action of which, instead of being general throughout the spectrum, will be limited to a particular part of it. I have now interposed a piece of red glass, which cuts off nearly all the light except the red; and now I interpose a piece of blue glass, that cuts off everything except the extreme violet. By introducing both these pieces in the beam, the spectrum is entirely obliterated.

In these latter cases we have instances, not of general, but of selective absorption, one substance cutting off everything but the red, and the other cutting off everything but the violet. Now the fact that we can absorb any definite part of the spectrum by properly tinted glasses provides us with a practical application of spectrum analysis in the manufacture of the coloured glass used for lighthouses or signals. Further, if astronomers could find a glass of a certain red, or a glass of a certain green colour, we should be able to see the solar prominences every day without a spectroscope.

FIG. 46.—Glass case for studying the absorption of liquids.

The first practical application which springs out of these phenomena of absorption is this, that as different substances are known by the effects which they produce on radiation, so also chemists find it perfectly easy to detect different substances by means of their absorption; for instance, the absorption spectrum of nitrous fumes can be shown by taking first our continuous spectrum, which we must always have to start with, and introducing some nitric peroxide between the source of light and the prism. The nitric oxide, immediately it comes in contact with the air, produces dense red fumes, and numbers of fine black lines will be seen immediately crossing the spectrum at right angles to its length, and to a certain extent resembling the solar spectrum with its Fraunhofer lines. Iodine is another substance which gives a coloured vapour, the absorption spectrum of which

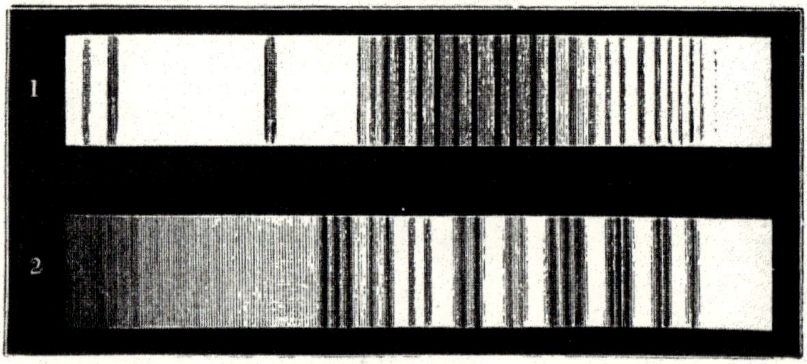

FIG. 47.—Absorption spectra of (1) iodine and (2) nitrous fumes.

is very definite and well defined. Fig. 47, Spectrum No. 1, shows the absorption spectrum of iodine vapour, and No. 2 that of nitrous fumes. We are not limited to these substances; we will try something else—blood, for instance, about which I shall have something more to say presently. We shall find that the action of blood upon the light is perfectly distinct from the action of those fumes which we have spoken of; and instead of having typical lines in the green and blue parts of the spectrum, we have two very obvious lines in the more luminous part of the spectrum. The colour of a solution of blood is not unlike the colour of a solution of magenta; but if, instead of using a solution of blood, we use a solution of magenta, we should have only a single black band (Fig. 48). The absorption spectrum of potassic

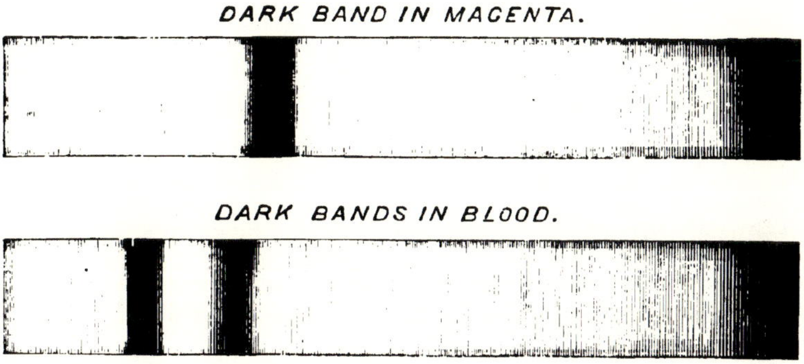

FIG. 48.

permanganate solution is another beautiful instance. We have here something totally unlike anything we have had before. Instead of the two dark bands which we saw in the case of the blood, or the single band in the case of magenta, we have four very definite absorption bands in the green part of the spectrum. So that you see the means of research spectrum analysis affords as far as regards radiation, is entirely reproduced in the case of absorption, and it is perfectly easy, by means of the absorption of different vapours and different substances held in solution, to determine not only what the absorbers really are, but to determine the presence of an extremely small quantity. Further, by allowing the light to pass through a greater thickness of the absorbing substance, the absorption lines are thickened and new regions of absorption are observed. This fact was discovered by Dr. Gladstone, who used hollow prisms containing the substance. I told you I had something more to say about the spectrum of blood, and this is not only an instance of the way in which the spectrum helps us in several important questions which, at first sight, do not seem at all connected with each other, but it shows the enormous power of research that is open to us. The colouring matter of blood, for instance, is found, like that of indigo, to exist in two perfectly different states, which give two perfectly different spectra. The colouring matter of blood is indeed capable of existing in two states of oxidation, which are distinguishable by a difference in colour, and also in their action on the spectrum. They may be made to pass one into the other by suitable oxidizing and reducing agents; they have been named by Professor Stokes, their discoverer, red and purple cruorine. Previous to the introduction of spectrum analysis, red and purple cruorine were perfectly unknown. Further, if by means of a spectrum microscope, such as I have already described, a blood-stain is examined, Mr. Sorby asserts that the thousandth part of a grain of blood—that is to say, a blood-spot so small that it only contains $\frac{1}{1000}$ of a grain, is perfectly easy of detection by means of this new method, and he has shown that its presence may be easily proved in stains that have been kept for a long time, and recognized even after a period of fifty years. He has also shown how it may be detected under the most unfavourable conditions, provided that a trace of hæmatin has escaped decomposition or removal; he has, in fact, successfully applied this method in several important criminal cases.

Another very interesting fact is, that when blood contains very small quantities of carbonic oxide gas in solution, it exhibits a very curious series of absorption bands. This fact is of considerable value in toxicological research, for in cases of poisoning by the so-called charcoal fumes, where, as is well known, the poisonous action is due to the formation of carbonic oxide, it can be readily detected by the peculiar bands which the blood under these circumstances exhibits.

Mr. Sorby has also applied the spectrum microscope to the study of blow-pipe beads, and has shown that in some cases as small a quantity as $\frac{1}{1000}$th of a grain of some substances can be thus recognized, even when mixed with other coloured bodies, which would interfere with the usual reactions dependent on colour alone.

In the case of radiation, as you know, we are able to determine the existence of new elements altogether. This is produced to a certain extent, as in the above case, in the absorption spectrum. Let me give you another practical application of this principle. Dr. Thudichum, as a result of researches made for the Medical Department of the Privy Council, has communicated to the Royal Society a paper in which he narrates the result of his inquiries on the yellow organic substances contained in animals and plants; and at the present moment it is impossible to say what important practical results may be expected as we come to know more about these substances.

especially in the matter of dyes, which I am sure is a thing that will commend itself to you.

Again, Mr. Sorby, in a communication to the Microscopical Society, brings the matter still nearer home. He shows us that, in the case of wines, he can, by means of the absorption bands, determine the very year even of vintage, and this, you will see at once, is a matter of very great importance. Let me read you an extract from one of Mr. Sorby's reports. He says:—"The difference for each year is at first so considerable that wines of different vintages could easily be distinguished; but after about six years, the difference is so small that it would be difficult or impossible to determine the age to within a single year. After twenty years, a difference of even ten years does not show any striking contrast, and the age could not, therefore, be determined to nearer than ten years by this process. However, up to six years, I think it quite possible to determine the age to within a single year. I took specimens of various ports from the casks, of different ages up to six or seven years, and labelled them in such a manner that I did not know the age of any, but could ascertain it afterwards by reference. I then made the experiments with great care, and found that, by proper attention to the details described above, I could correctly determine the year of vintage of each particular specimen." (*Chemical News*, December 17, 1869, p. 295.)

We have, in fact, a definite method of analysis of animal and vegetable colouring matter, and also of the colouring matter of decayed wood. Nor is this all, for, in another communication—for these things are now beginning to crowd upon us, and they will continue to do so much more by-and-by—Dr. Phipson asserts that this new method is perfectly competent to indicate any artificial coloration of wine. Mr. Sorby, on the other hand, has given his attention to beer; so that you see, if I have been taking you occasionally to the stars, I sometimes have the opportunity of travelling a great deal nearer home.

Mr. Sorby has also made some extremely delicate and interesting researches on the colouring matters existing in leaves. He has been able to identify numerous colouring principles, which he has arranged in five distinct groups: these groups rejoice in the names of chlorophyll, xanthophyll, erythrophyll, chrysophyll, and phaiophyll, the absorption spectra of which are perfectly distinct and well marked. It is found generally that leaves contain colours belonging to several groups, and frequently more than one of the same group. Mr. Sorby also finds that the change of colour which takes place in autumn consists chiefly in the disappearance of the chlorophyll, which renders the remaining colours visible, and these most frequently are of a yellowish tint. Some leaves, however, turn red in the autumn: this appears to be due to a falling off of the vital power of the plant, for by artificially diminishing the vital power, the intensity of this red colour is increased.

One great value of this method of research is that it enables us to recognize special colouring-matters, even when mixed with several others, and to determine the particular conditions in which they occur in plants or animals—whether in a solid state or in solution—and whether those dissolved out by reagents exist as such in the living organisms, or are the products of decompositions.

So that you see, on the whole, at the present moment, I think we may be full of hope that the new process may gradually lead to many more practical applications; but really we cannot say much about them at present, because the introduction of spectrum analysis is so recent. We are, however, already furnished with another instance of the close connection there always must be between any great advance in physical inquiry and the application of the skill of our opticians to aid us in the inquiry. We have the Sorby-Browning spectrum microscope, and at once a large number of people can study the beautiful phenomena which this new method of research has opened up to us, in regions where formerly it was almost impossible to imagine that science, or even the practical affairs of earth, should in any way benefit by such a study.

Having thus dealt very briefly with some of the more practical applications of the subject, I must now take you a somewhat distant journey to the sun and to the stars; and I must, in the first instance, attempt to connect the two perfectly distinct classes of phenomena which I have brought to your notice—the phenomena, namely, of radiation and the phenomena of absorption; and this connection between radiation and absorption is an instance of the slow growth of science. I remarked to you in the former lecture, that Fraunhofer, at the beginning of this century, had a very shrewd suspicion of the perfect coincidence of place in the spectrum between certain dark lines which he saw in the spectrum of the sun, which I promised to explain to you on this occasion, and the bright lines in the spectrum of sodium. You know how very simple the spectrum of sodium is: you will, perhaps, think it very strange indeed that such a simple thing was not explained very long ago. But Fraunhofer at the first suspected, and after him many of our greatest minds suspected, that there was some hidden, wondrously strange, connection between the double yellow line which you will remember is characteristic of sodium, and a certain double line which exists among the strange black lines of the solar spectrum, which I begged you to banish from your minds on the last occasion, when we were merely dealing with radiation. But now I must ask you to bear with me while I attempt to make clear to you all the strange facts concerning these black lines. I have been favoured by Dr. Gladstone with an extract from Dr. Brewster's note-book, dated St. Andrews, October 28th, 1841. In it Brewster says:—"I have this evening discovered the remarkable fact that, in the combustion of nitre upon charcoal, there are definite bright rays corresponding to the double lines of A and B, and the group of lines *a* in the space A B. The coincidence of two yellow rays with the two deficient ones at D, with the existence of definite bright rays in the nitre flame, not only at D but at A, *a* and B, is so extraordinary, that it indicates some regular connection between the two classes of phenomena." The double lines A and B refer to some of these dark Fraunhofer lines in the solar spectrum, which for convenience of reference were at first called after the letters of the alphabet; we now find that their number is so enormous that it is absolutely impossible to attempt to grapple with them in any such method, but these names are still retained.

The explanation of the coincidence between the two bright lines of burning sodium vapour and the two dark lines D in the solar spectrum was first given by Professor Stokes about 1852.

It is this. The light emitted by an incandescent vapour is due to the vibrations of its molecules, as a sound note emitted by a piano wire is due to the vibration of the wire. You have only to go into a room where there is a piano, and sing a note, to find that the wire which corresponds to your note will respond to your voice. Now, in the same way, when light is passing through a vapour the molecules of which vibrate at any particular rate, they will be urged into their own special rate of vibrations by the vibrations of the light which correspond to that particular rate which is passing through them. Hence the light will, so to speak, be sifted, and the force it has exercised in impelling the particles in the interrupting vapour to vibrate will tell upon it; and in this way, those particular vibrations which have had the work to do will be enfeebled.

It is clear that the parts of the spectrum thus reduced in brilliancy will depend upon the vapour through which the light has passed. If sodium vapour be traversed, then the light corresponding to the bright lines of sodium will be enfeebled.

This great law, to which the researches of Stokes

and Stewart and Ångström have led, and which has been established by the experiments of Foucault, Kirchhoff, and Bunsen, may be summed up as follows: *Gases and vapours, when relatively cool, absorb those rays which they themselves emit when incandescent;* the absorption is continuous or selective as the radiation is continuous or selective.

Now let me state to you how this discovery was finally established by Kirchhoff. In my notice of the spectroscope in the first lecture, I had so much to say that there were several details it was absolutely essential I should curtail. One of these details was the scale by which the positions of the different bright or dark lines which are seen in the different spectra are registered, so that we may say that such a line occupies such and such a position, and such another line occupies such another position, with regard to something else. When Kirchhoff and Bunsen, two German chemists, were engaged in mapping the spectra of the elements—a research which at its commencement had nothing whatever to do with the sun—they came across this difficulty of a scale. How could they get a good scale? I have already referred to some very obvious arrangements that might determine the actual position; for instance, the observing telescope may be made to move along a graduated arc, so that by moving the telescope for the different rays and fixing it when in a proper position to see a particular ray, you might read off the index placed on the arc to a great nicety by means of a graduated *vernier* working on the curve of the arc; or you may, by a modification of the instrument, use a reduced photographic picture of a scale, so that the thing to be measured and the actual scale would appear in the field of view at the same time. Kirchhoff and Bunsen tried these methods, but they did not like them. Then it suddenly struck them that, as they made their experiments in the day-time, they might use as a scale the black lines in the solar spectrum, which had not been known to change since the time of Wollaston, who discovered them. When working in the day-

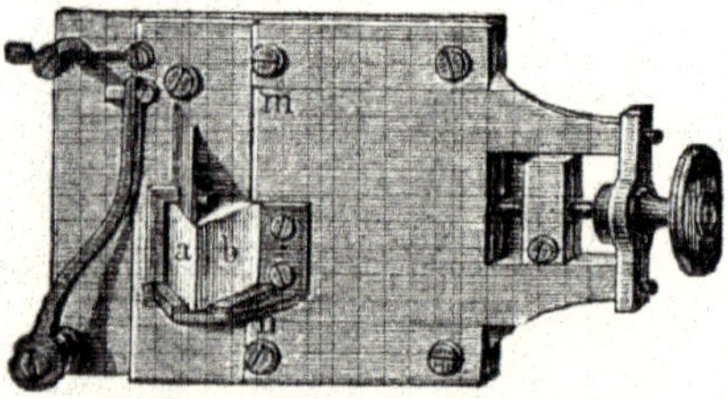

FIG. 49.—Steinheil's slit, showing reflecting prism.

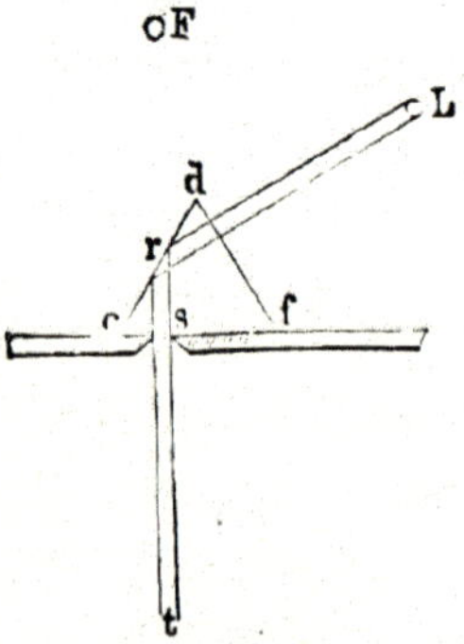

FIG. 50.—Path of light through reflecting prism and into the slit.

time, they had thus the solar spectrum visible in one half of the field of view of the telescope, which was easily managed by placing a reflecting prism over one half of the slit, as is shown in the enlarged slit in Fig. 49, so as to light one half of the slit by the sun, and the other half by whatever substance was under examination. With this arrangement they set to work with infinite care, and made a map of the solar spectrum. Such was their proposal: first, to map the unchangeable solar spectrum, and then, having this unchangeable scale, about which there could be no mistake, always visible, they would be able to refer to the dark lines in it all the unknown phenomena they were about to investigate in the bright lines of different vapours and gases. Having got this idea of the scale well into their minds, they were exceedingly anxious to test this question, which, as I have told you, was raised by Fraunhofer and many other men before them, of the asserted coincidence of the bright sodium line with the dark solar sodium lines: with a very delicate instrument, Professor Kirchhoff made the following remarkable experiment:— "In order," says Kirchhoff, for these are his own words, "to test in the most direct manner possible the frequently asserted fact of the coincidence of the sodium lines with the lines D"—(that is to say, of the bright double line of sodium in the yellow part of the spectrum, with the double line D of the solar spectrum) —"I obtained a tolerably bright solar spectrum, and brought a flame coloured by sodium vapour in front of the slit. I then saw the dark lines D change into bright ones." That is to say, in the spectrum of the sodium which was burning in the flame were bright lines so exactly coincident with the two dark lines in the solar spectrum, that the bright lines of the sodium spectrum put these dark lines out altogether, so that they seemed to vanish, as it were, from the solar spectrum. He goes on: "In order to find out the extent to which the intensity of the solar spectrum could be reduced without impairing the distinctness of the sodium lines, I allowed the full sunlight to shine through the sodium flame." Here he varies the experiment. In the first instance he used a very feeble beam of sunlight, but he now allows the whole glare of the sun to enter the slit. What was the result? "To my astonishment, I saw that the dark lines D appeared with an extraordinary degree of clearness." That is to say, the lines which came from the sodium in the first instance were sufficiently bright to entirely eradicate the dark lines from the solar spectrum, but the two lines D were now so utterly powerless compared with the light of the sun, that they actually appeared as black lines, and coincident with the two lines D in the solar spectrum.

We have seen that the bright line due to the radiation from sodium vapour can be very easily obtained by placing some sodium in a colourless gas flame, but if we now pass the continuous light coming from the carbon points of an electric light, or from the oxy-hydrogen lime-light, through this same sodium flame, the result will be that we obtain a black absorption line on a continuous spectrum, in precisely the same position as the yellow line was originally. This is Kirchhoff's crucial experiment, which at once determined not only that the dark line in the sun was absolutely coincident with the bright line of sodium vapour, but that, under certain conditions, bright incandescent sodium vapour could actually be made to absorb the light coming through it, and reverse its own spectrum. Kirchhoff goes on:—"I then exchanged the sunlight for the Drummond or oxy-hydrogen lime-light, which, like that of all incandescent solid and liquid bodies, gives a spectrum containing no dark lines." When this light was allowed to fall through a suitable flame, coloured by common salt (or chloride of sodium), dark lines were seen in the spectrum in the position of the sodium lines." You may imagine that this conclusive experiment—perhaps the most wonderful experiment that has been made during the century—gave Kirchhoff food for thought, and at once his genius travelled to a possible explanation of this strange fact he had observed; a fact, as you know, entirely in accordance with the previsions of Professor Stokes, Dr. Balfour Stewart, and Foucault. Kirchhoff said to himself, "I

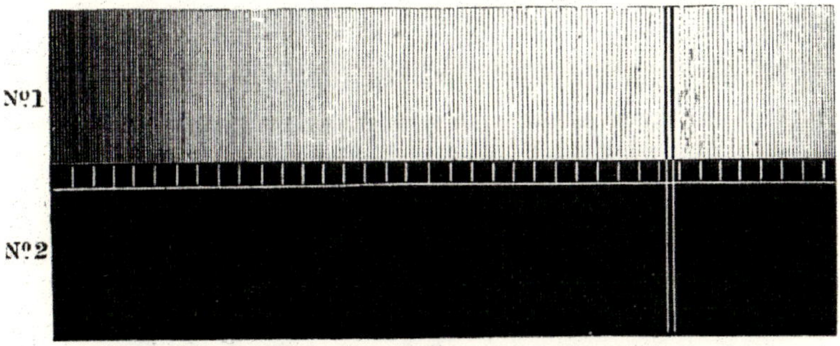

FIG 51.—Coincidence between the bright line given out by sodium vapour and the dark line produced by the absorption of sodium vapour.

have now got the bright lines in the spectrum of the vapour of sodium coincident with the two dark lines in the solar spectrum. What does it mean?" And again the philosopher was not at fault. He said to himself—it is almost possible to see the train of his reasoning in his memoirs—"Sodium has a most simple spectrum; suppose I take the most complicated spectrum I can find." He took for this purpose the spectrum of iron, which I think you will acknowledge to be one of sufficient complication, for the spectrum is traversed by lines throughout its whole length, and I may tell you at once that no less than 460 lines have been already mapped, and their positions are now thoroughly well known to us—as well known as the position of any star in the heavens. Kirchhoff tried the iron spectrum, and he found, absolutely corresponding in position in the spectrum and in width and darkness to the bright iron lines which he saw, black lines in the solar spectrum. He waited no longer; he instantly convinced himself, and soon convinced the world, that he had discovered this very remarkable fact, that gases and vapours have the power of absorbing those very rays which they themselves give out when in a state of incandescence. So

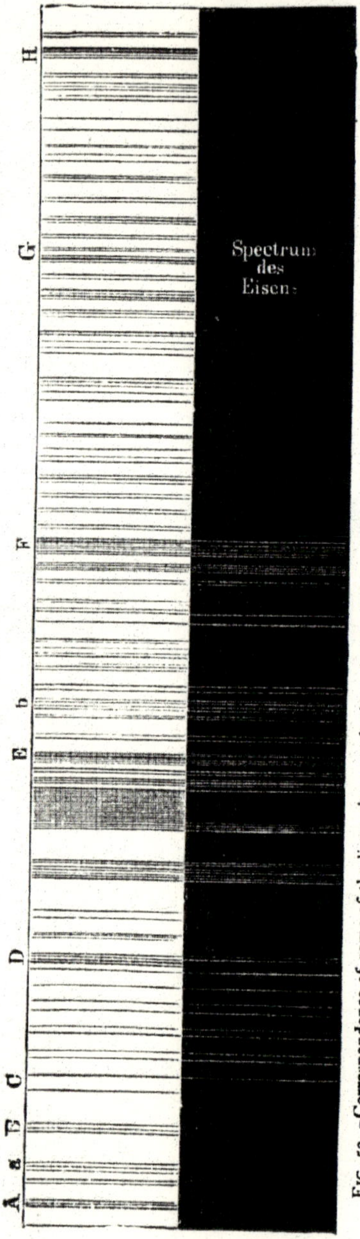

Fig. 52.—Correspondence of some of the lines given out by iron vapour (below), and of some of the Fraunhofer lines in the solar spectrum.

that, if you take sodium, and get its bright lines, and mark their positions on the screen, and then observe a continuous spectrum, and interpose sodium vapour in the path of the beam, you will find black lines absolutely corresponding with the bright ones; that is to say, that the sodium vapour has the faculty of entirely eating up, absorbing, or stopping that light which would otherwise go on to the screen. In the case of iron, it is worthy of notice that when Kirchhoff made his discovery, he was only able to obtain a spectrum of iron consisting of something like 90 lines, but since then the spectrum of iron has been mapped to the extent of 460 lines, and sure enough there are solar lines corresponding to nearly all the 460 bright lines which we are able to get in our laboratories. Not only was the bright line of sodium *reversed* or changed into a dark one, but it was soon found that the lines of other metals, such as lithium, potassium, strontium, calcium, and barium, could be reversed in a similar manner. This grand discovery of Kirchhoff's met with immediate acceptance, and with it you see at once the explanation of the wonderful black lines discovered by Wollaston, about which I said something in my first lecture. The riddle of the sun was read to a certain extent, and Kirchhoff read it in this way. He said:—"There is a solid or a liquid something in the sun, giving a continuous spectrum, and around this there are vapours of sodium, of iron, of calcium, of chromium, of barium, of magnesium, of nickel, of copper, of cobalt, and aluminium; all these are existing in an atmosphere, and are stopping out the sun's light. If the sun were not there, and if these things were observed in an incandescent state, we should get exactly these bright lines from them." Later researches by many distinguished physicists have shown that the following terrestrial elements are present in the vaporous condition round the sun:—

1. Sodium.	5. Iron.	9. Zinc.	13. Manganese.
2. Calcium.	6. Chromium.	10. Cadmium.	14. Aluminium.
3. Barium.	7. Nickel.	11. Cobalt.	15. Titanium.
4. Magnesium.	8. Copper.	12. Hydrogen.	

Kirchhoff further imagined that he had reason to believe that the visible sun, the sun which we see—and we may take the sun as an example of every star in the heavens—was liquid.

In the sun we have, first, a bright, shining orb, dimmed to a certain degree at the edge; and here and there, over the sun, we see what are called spots. Kirchhoff wished, not only to connect his discoveries with the solar atmosphere, but was anxious to connect them with this dimming near the limb and the spots. He said that the solar atmosphere, to which all the absorption lines were due, extended far outside the sun, and formed the corona; and that this dimming of the limb was really due to the greater absorption of this atmosphere, owing, of course, to the light of the sun travelling through a much greater length at the limb than at the centre of the disc. Furthermore, he said that the sun-spots, which astronomers, from the time of Wilson, had asserted to be cavities, were nothing but clouds floating in this atmosphere of vapour. Such was the very bold hypothesis put forward by Kirchhoff—an hypothesis which you see at once explains these strange observations from Wollaston upwards, including Fraunhofer's observations of the spectrum of the sun and stars, and the brilliant ideas of Professor Stokes, Dr. Balfour Stewart, and others in other lands. A little simple experiment, made by means of a little sodium vapour and a beam of sunlight, with the powerful aid of a little prism, gave us this tremendous knowledge about distant worlds, so immeasurably remote that it seemed absurd for men to try and grapple with any of the difficulties that are presented to us. Such, then, is Kirchhoff's theory of the sun, which I hope I have been able to make clear to you. There is a something—Kirchhoff said it was a liquid—which gives us a continuous spectrum, and between our eye and that incandescent liquid surface there is an enormous atmosphere, built up of vapours of sodium, iron, and so on; and the reason that we get these dark lines is, that the molecules of the substances named absorb certain rays, because when they are in an incandescent state they produce them. This brilliant idea of Kirchhoff's was soon carried, as you know, to the stars, by Mr. Huggins in our own country. In Fig. 35 will be seen the spectra of two stars, Aldebaran and α Orionis (Betelgeux), which are so distant that it is absolutely impossible to measure their distance from us. We know a great deal about our own sun, but these suns are so lost in the depths of space that it is quite impossible that we can get anything like a correct knowledge of their size, or know much of their belongings. By means of the prism, however, we learn in a moment a great deal. In the first star, we get three lines, due to the absorption of magnesium vapour, as we get them in the sun. We know, therefore, that magnesium vapour is present in the atmosphere around that sun (Aldebaran) in exactly the same way as round our own. We also get some of the iron lines, the lines of sodium, and the lines of hydrogen, calcium, and a few other elements—nine in all. At the base of the diagram you see indications of the elements, with the bright lines of which Mr. Huggins has compared the black lines which you see in the spectrum of these heavenly bodies. By means of the star spectroscope and of the induction coil, Mr. Huggins tested these lines, as Kirchhoff did in the case of the sun, by actually getting the vapour of magnesium visible at the same time in the spectroscope: and thus you see in a moment that there is no difficulty at all in determining their coincidence, when you have the two things brought so closely side by side. If I had time, I might remark on the presence of some elements here and the absence of others; but there is one remarkable fact about this lower star (*a* Orionis) which I must mention. As far as its spectrum goes, it appears that the gas hydrogen, which is a very important element in our sun's atmosphere, as we gather from the great distinctness of the hydrogen lines in the solar spectrum—and not only in our sun, but in a great many others—is absolutely absent, whilst magnesium, sodium, calcium, &c. are present.

So far, then, you see that this little prism has enabled us to read a great many secrets of the sun and of the more distant stars; and we must acknowledge that Stokes' and Kirchhoff's hypothesis is a very magnificent one, and we can but wish that there were more men like them, who, undismayed by the failure of those who for very nearly a century before their time had been endeavouring to unravel these secrets, were still prepared to go on, and endeavour to find them out by means of a prism and a simple sodium flame.

Now, astronomers—who, as I told you, from the time of Wilson had imagined that the sun-spots were cavities—very soon began to quarrel with this hypothesis of Kirchhoff's, who said that the sun-spots, instead of being cavities, were really clouds floating in the atmosphere. They remarked, and I think with truth, that to make such an assertion was altogether opposed to the evidence of the telescope. And I think I may say that the astronomers have now carried the day, for another line of independent research altogether—I mean the researches into the constitution of the sun by means of the spectroscope—has come to the aid of the astronomers, and it looks very much as if we must still hold to the opinion that Wilson in his observations, now more than a century old, was perfectly right, and that Kirchhoff's analysis, as far as it deals with the sun-spots, is susceptible of improvement. In the remarks I made in my former lecture on radiation in connection with the red prominences visible during eclipses, I drew your attention particularly to the hydrogen lines, and told you that the red flames are, for the most part, composed of hydrogen. There the prism comes to our aid in a very remarkable way indeed. It is clear to you, I think, after what I have said about absorption, that the darkening of the sun's surface, which we call a spot, is really a thing about which the prism can tell us a great deal. For instance, take a sun-spot, in which the usual brilliancy of the sun in the other parts of its disc is altogether wanting. There is not only great darkness here and there, but wonderful turnings and twistings and bendings of this solar envelope, which I have already told you Kirchhoff asserts to be a liquid one, but which I think a little consideration of Fig. 53 will show you is more probably gaseous, or cloudy, than liquid. It is obvious, I say, in this case that there was a great probability of the spectroscope being able to tell us something about this absence of light, for an absence of light means one of two things; it means either that there was a defect in radiation, or that there was some excess of absorption, and I may say that this difference—which I hope you now all thoroughly understand—really formed the battle-ground between the English and French astronomers until a few years ago. Long after Kirchhoff's experiment, M. Faye, a distinguished member of the Institute of France, went all over the work again, and declared that the sun-spot was dark, because we there got the light, not from the brightly shining envelope, but from some feebly radiating gas inside the sun; that the sun was a gigantic bubble, the bubble being nothing else than the photosphere—the liquid sphere of Kirchhoff—the interior being composed of gas, glowing at such an enormous temperature that the light we got from it was extremely feeble. You will see in a moment that, if the sun-spot were really due to the radiation from gas, we should get from that sun-spot a selective spectrum, that is to say, a spectrum with bright lines. The English astronomers said: "No; a sun-

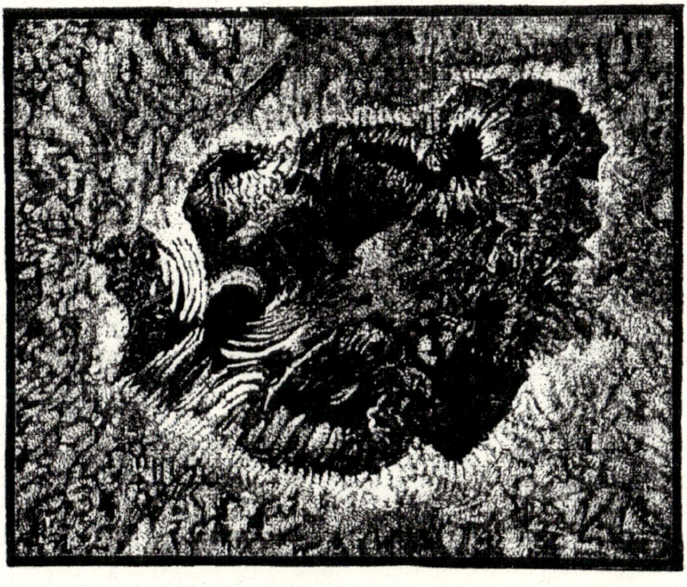

FIG. 53.—A sun-spot (Secchi), showing the "straws" in the penumbra, and the irregular masses on the general surface.

spot is not due to defective radiation at all; there is something over the bright portion of the sun which eats away the light:" whether the light was eaten away generally,—whether, in fact, we had an instance of general or selective absorption,—was not stated, but what they did distinctly state was, that the sun-spot was simply an indication of absorption. So that, you see, here was a thing which a spectroscope might settle almost at once, provided always that a good sun-spot could be obtained for the experiment. This was done in 1866. Fig. 54 gives an idea of what is seen when we observe a small sun-spot, and it is one which is full of meaning. Here is a very clear image of the solar spectrum near the double line D, and also the double D itself. If it were possible to have given you the whole of the sun's spectrum on the same scale as this, it would require an engraving yards in length, but it would be almost impossible to make my meaning clearer than I hope I can do by this small

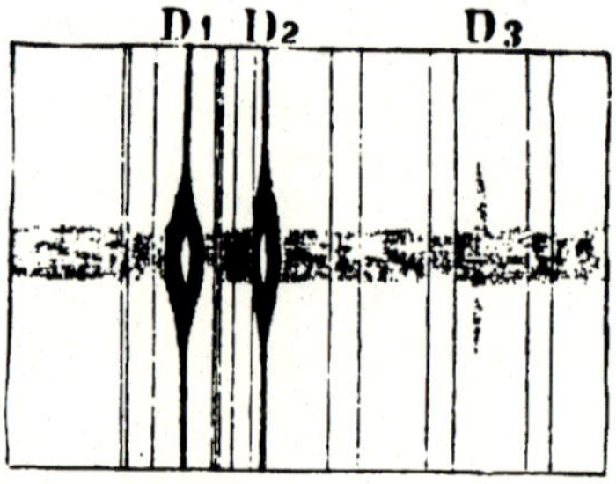

FIG. 54.—Spectrum of sun-spot. (Young.)

portion; and I must, therefore, ask you to take for granted, that the dark line which you see running along this yellow portion of the spectrum would really run along the whole length of the spectrum, from the extreme red to the extreme violet. This, then, you see in a moment, was an indication of general absorption; that is to say, in the way in which the light is affected by its passage through the prism, we have the problem settled in an instant, that a sun-spot is due to general absorption at all events. Further, in observing the spectra of different sun-spots, it was found that the spectrum of the middle of the sun-spot is much darker than the outside. So that you see this simple experiment tells us, not only that the sun-spot is due to general absorption, but that there is more general absorption in the middle of the spot than at its edge. This is the way in which this little prism is able to deal with these great problems.

But I have not yet done with this spot-spectrum. Not only is there this *general* absorption, but there are indications of increased *selective* absorption in the case of the line D, as I could also show if I were dealing with the iron lines, the magnesium lines, or the other well-known lines of the solar spectrum. Not only, then, have we a general absorption, increasing as the middle of the sun-spot is approached, but this sodium line D is also thickened, so that we have, as a result of a single examination of a single sun-spot, the fact that a sun-spot is due to *general* absorption, plus *special* absorption in some particular lines.

Now, in what I said some time since on the radiation of hydrogen, I pointed out to you that the F line of hydrogen was different from the C line—in fact, I showed that it widened out towards the sun—and I also told you that Dr. Frankland and myself have asserted that that widening out is due to pressure, and we have been able artificially to widen out this F line of hydrogen by increasing the pressure. Now it struck us that possibly we might find some connection between that widening out of the F line of hydrogen and the widening out of the sodium line in the spot which I have just shown you. There is an experiment by which it is perfectly easy for us to reproduce this artificially, so that you see we can begin at the very outside of the sun by means of hydrogen, and see the widening of the hydrogen lines as the sun is approached; and then we can take the very sun itself to pieces, and, by examining the pieces, see that the sodium lines vary in thickness in different parts of the spot, as the hydrogen does outside the spot region altogether: in fact, the pressure is continually increasing down in the spot exactly in the same way as it increases in the hydrogen envelope towards the sun.

If we take a tube containing some metallic sodium sealed up in hydrogen, and pass a beam of light from the electric lamp through it; by decomposing this beam with our prisms we shall obtain an ordinary continuous spectrum without either bright or dark lines, but by heating the metallic sodium in the tube which is placed in front of the slit, we really fill that tube with the vapour of sodium; and as the heating will be slow, the sodium vapour will rise very gently from the metal at the bottom, so that we shall get layers of different densities of sodium vapour filling the tube. Immediately the sodium begins to rise in vapour, a black absorption line shows itself in our spectrum in precisely the same position as the yellow line of sodium, and you will find that the thickness of the sodium absorption line will vary with the density of the stratum of vapour through which it passes. Thus from the upper part of the tube we obtain a fine delicate line, which gradually thickens as we approach the bottom of the tube, and thus we produce the appearance in the spectrum of the spot where the layers of sodium vapour are very dense, and the very fine delicate line of the sodium vapour when thrown up into the sun's chromosphere.

We must next speak of what happens in the case of the magnesium lines. A very obvious magnesium line is lettered *b* in the solar spectrum. It is a triple line, separated by different intervals, shown in the frontispiece. There is a very important fact connected with these lines, which appear when magnesium vapour is thrown up into the envelope which I have called the Chromosphere. By means of the new method of research, it is quite possible to see, as I explained to you on a former occasion, what passes, which the eye could not possibly see. For instance, it is quite possible, by means of the spectroscope, to detect the existence of magnesium vapour outside the sun, although you know that, except during eclipses, we are never able to see these vapours. What I wish to call your attention to in the present case is this. We have there the three magnesium lines, and two of them are much thicker than the remaining one: and these two lines travel very much higher into the outside region than does the third one. Now, you will see in a moment that that indicates to us a fact something like this,—that the spectrum of magnesium, such as is generally at work, which cuts out these very black absorption lines in the solar spectrum, while the sodium gives us the yellow line D, is really a thing which is competent to give us three lines. This vapour, I say, is a thing, generally speaking, competent to give us three lines in this position;

but if it so happens that when the magnesium is thrown up to a particular height we simply get two lines, the third stopping short, I think you will see that there is some force in one's reasoning, when one suggests that possibly in those regions where we find the hydrogen F line thin instead of thick, as I have shown it to you, and where the magnesium lines become reduced to two instead of three, the spectrum of magnesium vapour, like the spectrum of hydrogen, becomes very much more simple by *the reduction of pressure*, and therefore, that we should be able artificially, as in the case of hydrogen, and as in the case of sodium, to reproduce this result. In fact, it is perfectly easy to reproduce it, for we find by reducing the pressure of magnesium vapour we really can reduce that triple line of magnesium to a double one ; so that, you see, we have three distinct lines of research, all leading us to the fact that where Kirchhoff placed an immensely dense atmosphere around a liquid sun, we really have vapour of considerable tenuity, by no means so dense as he supposed.

There is another point of very great interest which I should bring before you.

Mr. Huggins, who has done so much in his researches on stars, told us some few years ago that the spectrum of that wonderful variable star τ Coronæ, which had been just discovered, indicated that, over and above the light which we got from the star generally, we get evidence of incandescent hydrogen in the spectrum, so that the spectrum was a thing such as had never been seen before ; for we got, in addition to the ordinary evidence of absorption visible in the spectrum of a star, as in the spectrum of the sun, indications also of selective radiation. There are indications of bright lines superposed above the others. Now, let me tell you—and this is a very important part of the question—that by observing the various changes that take place in our central luminary, it is quite possible to see on the sun almost any day evidence of its being violently agitated; that there are certain regions of the sun which appear exactly as that variable star did—that is to say, in addition to the ordinary absorption lines visible in the solar spectrum, the spectrum of these regions indicates to us that the hydrogen, instead of being

FIG. 55.—Spectrum of τ Coronæ. (Huggins.)

black, instead of reversing the spectrum, as you have seen it in these spectra that I have shown you, really is bright, or else the hydrogen lines cease to be visible altogether, as in *α* Orionis.

I have to give you, as the last application of spectrum analysis, the power which the prism gives us of investigating, so to speak, the meteorology of the sun, the velocity with which the different stars are moving through space, and the velocity with which the storms are travelling over the face of our central luminary. Many of you know, no doubt, that Mr. Huggins, in his observations of the spectrum of the star Sirius, saw that the hydrogen lines were much developed; and in a further examination, carried on by the method in which the spectrum of hydrogen and other vapours which he wished to examine were absolutely visible in the field of view at the same time as was the spectrum of the star, Mr. Huggins was astonished to find that the hydrogen lines no longer occupied their usual positions, but that they were all jerked, so to speak, a little to the side of the place which they occupied in the spectrum of the hydrogen which he rendered incandescent in his tubes. The F line of hydrogen which he observed in the spectrum of Sirius he found did not exactly occupy the same position in the spectrum as did the

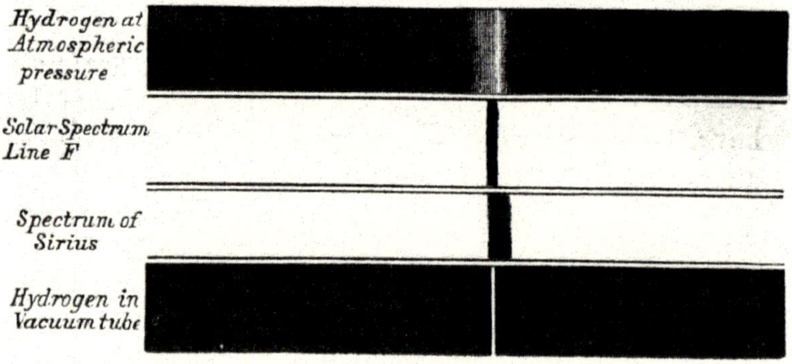

FIG. 56.—Alteration of wave-length of the hydrogen in the atmosphere of Sirius.

actual F line of hydrogen, the incandescent hydrogen with which he compared it (Fig. 57). Owing to a physical law, which I have not time to explain to you now, it is perfectly easy, by means of the prism, to determine the velocity with which the light-source is moving to or from us; and therefore, if this holds good for absorption, we could determine the velocity with which any absorbing medium is rushing to or receding from us. In the case of Sirius, for instance, Mr. Huggins determined that the velocity of the star in a direction from the eye, the measure of recession, was something like twenty miles a second. I am sorry I have not time to fully explain this very beautiful adaptation of the spectroscope, but I may say that the position of a line, bright or dark, in the spectrum depends upon its wave-length—that is to say, the length of the wave of light which produces that colour. Thus, the length of a wave of red light is about $\frac{1}{39000}$ of an inch, and that of a wave of violet light is about $\frac{1}{57500}$ of an inch. I think when I mention that, you will see at once the possibility of determining any alteration of velocity—for an alteration of wave velocity we have, or appear to have, whether we move towards an object, or whether an object moves towards us, just in the same way as in the case of sound, and in the case of a wave reaching the shore. Suppose yourself a swimmer carried on a wave; if you are going with the wave it seems long, but if you attempt to swim against it, it seems short. So with all these waves, beating from all these orbs peopling the depths of space on to the earth. If by the motion of those bodies, or by our own motion, the waves are crushed together, we get an alteration in the light, which the prism alone is able to determine. If the luminous object is approaching the eye rapidly, the vibrations causing light will, of course, fall on the eye more frequently in the same time than if the bodies were at rest—or, in other words, the waves will be shortened; then the position of the dark or bright lines, as the case may be, will be shifted in the direction of the most refrangible rays—that is to say, towards the violet; whilst if the bodies are separating, the shifting will take place in the direction of the red or least refrangible rays. In the case of Sirius the star was receding from us, and we got longer waves, and the lines are nearer the red end of the spectrum to such an extent as to leave unaccounted for a motion of recession from our sun amounting to something between 18 and 22 miles per second. Other stars, such as Betelgeux, Rigel, Castor, Regulus, and many of the stars in Ursa Major, are found to be moving away from the sun. Some, however, move rapidly towards us. Arcturus approaches us with a velocity of 55 miles per second; Vega and *a* Cygni, Pollux and *a* Ursa Majoris, also approach the sun with a velocity varying from 40 to 60 miles per

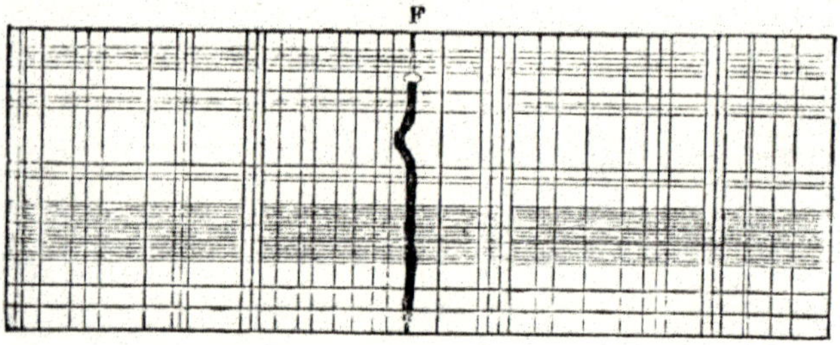

FIG. 57.—Deviation of the F line in a spot-spectrum.

second. If now we take a spot-spectrum (Fig. 57), in which instead of the sodium line D we have the F line of hydrogen, this strange crookedness which you notice is really a crookedness due to the fact that in one place we have incandescent hydrogen rising up with tremendous velocity, and in another we have it rushing down cool with tremendous velocity; again, we have hydrogen in a different condition altogether. We know that in this case we have a variation of velocity, because we get distinct changes in one direction or the other, and we get changes in both directions. We can determine by the amount of crookedness of the hydrogen, whether bright or dark, how far it is driven from its normal condition, and then how fast per second the hydrogen is travelling. In one case the velocity was something like 38 miles a second; in other words, we had heated hydrogen coming up at the rate of something like 38 miles a second, and cool hydrogen rushing down at something like an equivalent rate. Now, we are not only enabled, by a practical application of the prism, to determine these up and down rushes on the sun, by

FIG. 58.—Shifting of the F line in a solar cyclone.

which we are enabled to learn much of its physical constitution, but also the rate at which storms travel over the sun—what we should call winds. The way that has been done will be perfectly clear on an inspection of the engraving (Fig. 58). It may appear strange to you that we should be able to observe a cyclone on the sun, but I hope to be able to prove to you that this is really a cyclone. Here is a spectrum of the region of the sun near the limb, and here is the hydrogen line. It is clear, if what I have said is true, that the incandescent hydrogen is there receding from us because the line inclines to the red. It is evident also, that in this case, when we get the line widened out towards the violet, it is coming towards us; therefore we have the thing travelling in both directions. It is obvious to you, I think, that if the slit enabled us to take in the whole cyclone, we should get an indication of motion in two directions; we should have the line diverted both towards the violet part of the spectrum, in the case of the hydrogen rushing towards us, and towards the red in the case of the hydrogen rushing away from us in this circular storm; and the extreme velocity will be determined by the extreme limit to which the hydrogen line extends. In this case, the storm was moving with a velocity of something like 100 miles a second, which, I dare say, strikes you as something terrible; but if you compare the size of the sun with that of the earth, I think you will see it was nothing very wonderful after all.

In further evidence of the truth of this, the last application of the spectroscope, I will show you two pictures of solar prominences 27,000 miles high, drawn at an interval of ten minutes. Here you see, first, the prominence as it appeared at a particular time on a particular day in March 1869 (Fig. 59). I wish to call your attention to the left-hand portion of the prominence, which you see is pretty straight. In

FIG. 59.—Prominence observed 14th March 1869, 11h. 5m.

ten minutes afterwards the whole thing changed, and, as you see by the next picture (Fig. 60), the nearly straight portion is quite gone. That will give you some idea of the indications which the spectroscope reveals to us of the enormous forces at work in the sun, merely as representing the stars; for everything we have to say about the sun, the prism tells us—and it was the first to tell us—we must assume to be said about the stars. And I have little doubt that, as time rolls on, not only will the spectroscope become, in fact, almost the pocket companion of every one amongst us, but it is utterly impossible to foresee

FIG. 60.—The same prominence, 11h. 15m.

what depths of space will not in time be gauged and completely investigated by this new method of research.

THE END.

Spotted something that isn't quite right?

Let us know and we'll fix it for future reprints.
And as a thank you for your diligence, we'll
send you a free Sothis book of your choice.

publisher@sothis.press

This page intentionally left blank

Made in United States
North Haven, CT
06 November 2026

21878605R00079